Chemistry and Science Fiction

Jack H. Stocker, Editor
University of New Orleans

American Chemical Society
Washington, DC

Library of Congress Cataloging-in-Publication Data
Chemistry and science fiction / Jack H. Stocker, editor.

p.   cm.

Includes bibliographical references and index.

ISBN 0–8412–3248–2

1. Science fiction, American—History and criticism. 2. Science fiction, English—History and criticism. 3. Science fiction films—History and criticism. 4. Science fiction—Study and teaching. 5. Chemistry in literature. 6. Literature and science. I. Stocker, Jack H., 1924–   .

PS374.S35C48   1998
813'.0876209356—dc21   98–4720
                                                                                CIP

The paper used in this publication meets the minimum requirements of American National Standard for Information Sciences—Permanence of Paper for Printed Library Materials, ANSI Z39.48-1984.

Dedicated to my wife,
Catherine Wolters Stocker,
and to our children,
Daniel and David.

Contents

Preface vii

Some (Extra-Terrestrial) Acknowledgments ix

Contributors xi

Introduction xix
James Gunn

An Overview

1. A Science Fiction Primer for the Uninitiated 3
Jack H. Stocker

2. Science in Science Fiction: A Writer's Perspective 21
Connie Willis

History and Tradition

3. The Search for a Definition of Mankind: H. G. Wells and
His Predecessors 35
Ben B. Chastain

4. Planetary Chemistry in 100 Years of Science Fiction 53
Mark A. Nanny

5. Beryllium, Thiotimoline, and Pâté de Fois Gras:
Chemistry in the Science Fiction of Isaac Asimov 77
Ben B. Chastain

6. The Improbable Properties of Imipolex G: Chemistry and
   Materials Science in Thomas Pynchon's *Gravity's Rainbow* . . .91
   Carl Trindle

7. Sherlock Holmes: The Eccentric Chemist 105
   James F. O'Brien

8. The Right Environment: Using Real Chemistry as a Basis
   for Science Fiction 127
   Harry E. Pence

9. On the Covers of Science Fiction Magazines 137
   Jack H. Stocker

The New Media: Television and the Movies

10. Where No One Has Gone Before: Chemistry in *Star Trek* 157
    Natalie Foster

11. Chemistry in TV Science Fiction: *Star Trek* and *Dr. Who* 173
    Penny A. Chaloner

Chemists at Play

12. The Endochronic Properties of Resublimated Thiotimoline . .205
    Isaac Asimov

13. Pâté de Foie Gras 211
    Isaac Asimov

14. Temporal Chirality: The Burgenstock Communication 227
    Michael J. S. Dewar

Encouraging Creativity in the Classroom

15. Science Fiction: A Classroom Resource 233
    Jack H. Stocker

16. Using Science Fiction To Help Teach Science:
    A Survey of Chemists and Physicists .241
    Clarence J. Murphy, Mary Ann Mogus, and Patricia M. Crotty

17. Space, Time, and Education 251
    John E. Arnold

Appendix: Recommendations for Further Reading 269

Index 285

Preface

In April, 1992, at the Spring Meeting of the American Chemical Society in San Francisco, the History of Chemistry Division offered a symposium entitled "Chemistry and Science Fiction." Scheduled in a rather small room accommodating perhaps 35 people, the first two talks were presented to an overflowing, standing-room-only audience while additional would-be attendees languished outside. Transferred to a much larger room, seating perhaps 100, even that capacity was taxed with a number of standees for the rest of the day's symposium. This book is an outgrowth of that highly successful symposium. For various reasons, some presentations could not be included, while some chapters reflect augmented presentations, and several chapters have been added. The difficulty in establishing clear boundaries for the science fiction genre is discussed in the first chapter of this book. Suffice it to say here that eligibility for inclusion in the symposium was deliberately set as broadly as possible and an occasional entry may raise an eyebrow or two. We can only request the tolerance of the reader who is quite certain his or her narrower definition of the field is a more accurate one.

The real difficulties came, however, with the "chemistry" part of the title. This is partly because stories are not textbooks, and the chemistry component may be a single large one (for instance, establishing that the nonsurvival of a planet colony is due to the trace concentration of beryllium in its soil, as appears in a novelette by Isaac Asimov) or several smaller ones (such as occurs in stories like *The Andromeda Strain* by Michael Crichton). Chemistry is not one of the sciences dominating the field of science fiction; quite likely physics, with its gadgets, laws, and grandiose concepts, vies with the social sciences, with their corresponding utopias, dystopias, and alien cultures, for that honor. The biological sciences follow a short distance behind. It can be suggested that perhaps the two most widely chemistry-oriented contributions lie in the biochemical and materials science domains. Examples of both of these areas are reflected in the following chapters. A lengthy listing of suggestions for further reading, often accompanied by supportive commentary, is provided in the book's concluding section.

An additional comment or two on what this book does and does not intend to do would be in order. While a number of authors enjoy pointing out and sharing with the reader examples of chemical bloopers, some of which are quite colossal, it was not the primary purpose of either the symposium or this book to take a totally analytical stance and critique the science fiction field. Rather, the goals were to share an enthusiasm, to make a few recommendations, and to persuade a few readers to stretch their minds and think in some unorthodox categories.

JACK H. STOCKER
University of New Orleans

Some (Extra-Terrestrial) Acknowledgments

The following terrestrials supplied the editor and his project with that extra amount of inspiration, encouragement, and support that demands recognition. Herewith, he expresses his acknowledgment and gratitude for the extra-terrestrial help:

★ To Jim Bohning, who began the planning for the original American Chemical Society symposium of the same name but found it necessary to withdraw his leadership from the project due to excessive demands on his time.

★ To those participants in the original symposium who were unable, for various reasons, to be included in this book: Bill Jensen, Naola Van Orden, and David Katz.

★ To those overburdened office secretaries in the Department of Chemistry of the University of New Orleans who typed and typed and $(typed)_n$: Keith, Mildred, Janice, and Jovian.

★ To Professor Waldemar Adam, of the University of Würzburg, who answered the question of whether the contribution of Michael Dewar had been published previously in copyrighted form. By diligent investigation, he determined that it had been published in a German translation and not in the English in which it had originally been composed.

★ To HIST, the ACS Division of the History of Chemistry, for its willing-ness to share whatever royalties might be realized from the sale of this book. (The original symposium was a result of its sponsorship.)

★ To Carol Emshwiller, the widow of the fondly remembered illustrator who signed his artwork EMSH, for her kind permission to include, roy-alty-free, a full-color duplication of one of his science fiction magazine covers. To Kelly Freas, one of the field's most admired artists/illustra-tors, for his generous permission to include, royalty-free, his multiple adornments that accompanied one of the Isaac Asimov spoofs in its original appearance in *Astounding Science Fiction Magazine*.

★ To George Schumacher, who manages copyright matters for Dell Pub-lishing Corporation, for his considerable help in obtaining permission to reproduce a number of copyrighted materials, as well as his courtesy and patient guidance to those of us unsophisticated in such matters.

★ To Kathy Strum and Cheryl Wurzbacher, truly caring copy editors and production coordinators, who arrived on board late but provided some very helpful guidance.

★ And, most particularly, to this book's ACS liaison and patron saint, Anne Wilson, for her professional skills and steady commitment to the project. That unwavering faith saw us through some very rough times. Thank you, Anne.

Contributors

John E. Arnold
As his science-fact article reports, Arnold was teaching in the department of engineering at Massachusetts Institute of Technology when his article appeared. He left that university for a position at Stanford University in the late 1950s. He passed away during a visit to Europe in 1964.

Penny A. Chaloner
Chaloner is a senior lecturer in chemistry at the University of Sussex in the United Kingdom, where she has been since 1983. She received her Ph.D. from Cambridge, pursued post-doctoral studies at Oxford, and held the position of assistant professor at Rutgers University prior to her present affiliation. Her particular research interests led to authorship of the *Handbook of Coordination Catalysis in Organic Chemistry* and co-authorship of the book, *Homogeneous Hydrogenation*.

Chaloner reports that she teaches a course in science fiction at Sussex, and she admits to having been a "Trekkie" since the 1960s, as well as a (Dr.) "Whovian."

She can be contacted at her university.

Ben B. Chastain
A native Alabamian, Chastain received his undergraduate degrees from Birmingham Southern College and his Ph.D. from Columbia University.

He has been on the faculty of Samford University for an impressive 35 years, serving as the chair of its department of chemistry since 1984. An active member of the American Chemical Society's Division of the History of Chemistry, he served as its chair in 1992 and a division councilor since 1996.

Chastain reports that his interest in science fiction began in his high school days, when he encountered the Healy-McComas anthology, *Adventures in Time and Space*. (Still the single best collection of Golden Age science fiction, it has been regularly reprinted since its first appearance in 1946—Editor).

Michael J. S. Dewar

Professor Michael J. S. Dewar passed away on October 10, 1997. One of the giants in the field of physical–organic chemistry, he is remembered for his sense of humor and his delight in challenging professional icons. The editor acknowledges a warm and rewarding correspondence with Professor Dewar for more than 40 years.

Natalie Foster

Foster is an associate professor of chemistry at Lehigh University in Bethlehem, Pennsylvania, where she is currently studying poly(vinyl alcohol) gels by NMR as part of a larger interest in the influence of intermolecular interactions on relaxations. She is also co-editing a freshman chemistry textbook.

Far more impressive, however, is Foster's startling disclosure, which she now reveals for the first time: Time travel *will definitely be a reality*. (She admits, however, that this will have the dire consequence of forever confusing verb tenses for generations of English scholars.) It seems that she has served/will serve in the 24th century aboard the U.S.S. *Enterprise* on its continuing mission under the command of her mentor and *very close personal friend* Captain Jean-Luc Picard. She adds that she is undergoing/has undergone special Federation training practice and, for reasons that she cannot disclose under provisions of the Prime Directive, she is pursuing/has pursued intensive anthropological inquiries into the politico-cultural role of dining-out in restaurants featuring excellent wine cellars. She welcomes/welcomed suggestions and support for this study, since clearly the future of the Universe is at steak/stake.

She may be contacted (at the reader's risk) at her university.

James Gunn

Gunn, currently Professor Emeritus of English at the University of Kansas, is regarded by knowledgeable SF readers to be an elder statesman of the field. Born in Kansas City, Missouri, Gunn spent three years in the U.S. Navy during World War II. He received his B.S. degree in journalism in 1947 and his M.A. in English in 1951, both from the University of Kansas.

There is essentially no aspect of science fiction to which he has not contributed creatively as well as notably. Among other activities, he has written 80 stories and 19 books, as well as plays, screenplays, radio scripts, verse, and criticism. Four of his stories were dramatized on NBC radio and one on Dealer Playhouse; *The Immortals* was televised on ABC-TV as "Movie of the Week" and later became the hour-long series, "The Immortal." In addition, he has edited seven books.

Gunn has lectured in 10 foreign countries for the U.S. Information Agency, and he has served as guest of honor at many U.S. and foreign SF conventions. Among his awards: the Eaton Award for Lifetime Achievement and a Hugo award for his nonfiction book about Asimov. Over the years, he has served as a focal point for SF archival matters, and he has organized meetings and short courses dealing with all aspects of science fiction.

He can be reached at his university.

Clarence J. Murphy

Murphy was born in Manchester, New Hampshire. He earned a B.S. and an M.S. in chemistry at the University of New Hampshire and then went west to the deeper snow of Buffalo, New York, where he obtained a Ph.D. in chemistry and a chemist wife at the State University of New York. His three daughters diversified the family professions by earning degrees in chemistry, family studies, and history.

Murphy has been a professor of chemistry at East Stroudsburg University in Pennsylvania since 1969, where he was the first chair of that department and the former acting dean of arts and sciences. He has held teaching or research positions at Ithaca College, St. Anselm College, Massachusetts Institute of Technology, and Cornell University. In addition, he has been a visiting research scientist in the Material Research Center at Lehigh University. He was trained as an organometallic chemist, and his research interests have centered in the field of polymer science with particular emphasis on phase changes and related properties.

Murphy reports that his first memory of science fiction is avidly reading a Buck Rogers comic strip in the Sunday paper "funnies." He also devoured

Jules Verne and H. G. Wells novels. He admits to being a fan of science fiction movies, particularly the awful classics such as *Attack of the Killer Tomatoes*. These, he points out, have a lot in common with reading and grading their modern science fiction analogues, often referred to as "laboratory reports."

He can be contacted at his university.

Mark A. Nanny

Nanny is an assistant professor of environmental chemistry in the School of Civil Engineering and Environmental Science at the University of Oklahoma. His research interests include the study of abiotic and biodegradation of pollutants in soils, sediments, and aquatic systems; the interaction of pollutants and metabolites with natural organic matter; and the characterization of complex organic mixtures, such as landfill leachate, among other related phenomena.

Nanny admits that he has enjoyed science fiction for as long as he can remember. Favorite authors include Herbert, Asimov and Clark, with the *Dune* series by Herbert and the *2001* series by Clark being particularly admired. He vividly recalls seeing the movie *2001* when it was first released, even though he was then but six years old!

He can be contacted at his university.

James F. O'Brien

O'Brien is a professor of chemistry at Southwest Missouri State University, where his research interests lie in carrying out theoretical calculations of the properties of inorganic molecules. The university has recognized his excellence both in teaching and research, naming him a University Distinguished Scholar in 1996.

O'Brien reports that his interest in his chapter's topic goes back to his days as a teenager when his father demanded that he read *The Hound of the Baskervilles*. He has been hooked ever since. He recalls with particular pleasure teaching "The Scientific Sherlock Holmes" to some 30 students in London and delivering a related talk on a number of occasions as an American Chemical Society tour speaker.

He can be contacted at his university.

Harry E. Pence

Pence (rumored to be known to his students as "Uncle Harry") is a distinguished teaching professor at State University of New York at Oneonta. His main research interest is the application of educational technology to chemistry teaching, expressed by organizing a number of symposia in this area. He has been particularly active in the American Chemical Society's Division of Chemical Education's Committee on Computers in Chemical Education. He won the SUNY Excellence in Teaching Award in 1987. He also wrote the study guides for the first three editions of *Chemistry and Chemical Reactivity* by John Kotz, et al.

Pence reports that he began reading science fiction in high school. During the late 1940s and throughout the 1950s, he read as much science fiction as he could get his hands on, including all the magazines: *Astounding*, *Galaxy* (from Vol. 1, No. 1!), and *Other Worlds*, as well as many others. Writing his chapter, he says, has given him an excuse to recall some happy memories.

He can be contacted at his university.

Jack H. Stocker

In 1991, after 35 years of teaching and research, including helping to found a new university, Stocker became Professor Emeritus of chemistry at the University of New Orleans. He maintains his university presence and continues to be professionally active in the area of organic chemical nomenclature (IUPAC Commission activities), serving as an ACS local section speaker (over 100 talks) and representing his local section in national ACS government as its councilor for more than 20 years.

Stocker reports that his life has always had a science fiction interface. At age 9, he "published" individually printed copies of a neighborhood newspaper that included his ongoing, badly drawn comic strip "Silly and Billy Go to Mars," where the characters met giants and other Buck Rogers menaces. He subsequently started a Big Little Book entitled *Dart Benson and His Rocket Ship*, wrote (and colored) over 50 pages of a new Oz book in which Dorothy reached Oz as a result of parachuting from a faltering airplane, and wrote a letter to the readers column of the magazine *Thrilling Wonder Stories*. All of which indicated "fannish" activities before 1940 and entitled him to membership in the science fiction organization "First Fandom."

Related, occasionally more adult, activities have continued since then. Stocker has been a guest at a number of local and regional science fiction conventions and has served as a volunteer/consultant/dealer's room purveyor of paperbacks at all the more accessible ones.

One very special occasion: In 1966, Stocker was invited by Louisiana State University in Baton Rouge to organize the first of a multiweek program dedicated to lectures and panels featuring appropriate experts, such as Buckminster Fuller, who would talk about the world of the future. The first program, however, was a panel discussion (via a multiple telephone hookup and well-placed amplifying speakers) among distant participants Isaac Asimov, John Campbell, Poul Anderson and Frederik Pohl. Stocker, on an empty stage before an audience of several hundred, served as moderator. The participants were introduced, responded to questions from him, and vigorously debated with each other.

He can be contacted at his university.

Carl Trindle

Trindle received his initial training in chemistry at Grinnell College in Iowa and obtained his doctorate from Tufts University. Currently, he is associate professor of chemistry and director of studies at Brown College (a part of the University of Virginia), where he teaches quantum chemistry to graduate students and the foundations of the chemical world to undergraduates. His research activities involve the study of electronic structures and reactions of open-shell systems—radicals and carbenes—by computational modeling. He is an ACS tour lecturer who writes for and speaks to general audiences on science in literature and scientific ideas as they are expressed in the visual arts.

Trindle reports that his short stays at the Indian Institute of Technology (Bombay), the Institute Rudjer Boskovic (Zagreb), the Israel Institute of Technology (Haifa), and the Middle East Technical University (Ankara) underscored for him the unifying power of science.

He may be contacted at his college, which offers discussions and courses in science fiction.

Connie Willis

Willis is a Colorado native and lives in Greeley, Colorado, with her husband (an associate professor of physics with a Ph.D. in science education who was part of a Colorado writing team that prepared three modules for the ACS Source Book) and their daughter, plus a bulldog, and two cats. In addition to her chapter, she also contributed a significant portion of the recommended reading section at the end of this book.

A list of Willis's awards, her guest-of-honor invitations to science fiction conventions, her toastmasterships at prestigious award ceremonies, and her continuing prodigious output of first-class stories is awesome. Highlights of her publications include *Doomsday Book, Lincoln's Dreams, Bellwether, Uncharted Territory*, and her latest, *To Say Nothing of the Dog*. She has won six Nebula awards and is the only SF writer ever to have won in all four Nebula categories. She has won six Hugo awards, the John W. Campbell award for best SF novel (*Lincoln's Dreams*), and six Locus awards, the latest being for *Bellwether*. (Locus is a generally recognized clearinghouse publication for nationwide SF activity). She is working on a new book, *Cape Race*.

She can be contacted by e-mail: cwwilli@bentley.unco.edu.

Introduction

What you hold in your hands is a welcome volume in which one of the physical sciences, chemistry, pays tribute to the literature of science—that is, science fiction. For almost the entirety of science fiction's raffish history, the sciences that SF has loved so much have turned their backs on it, while at the same time nursing for SF a private vice, a secret passion. Now perhaps, the sciences are coming out of the closet, and chemistry, like physics and astronomy, may be willing to admit that, yes, an attraction for science fiction is all right, maybe even worthwhile.

At least the pioneer symposium at the 1992 American Chemical Society meeting suggests that springtime may be at hand. This collection of essays is another sign of reconciliation. And about time. Although science fiction is not literally a fiction about science, it is about the same things science is about—they share basic beliefs: that the universe is knowable, that everything that lives, including humans, is the product of evolution and adapts to its environment, that humans and other rational creatures can understand this fact and choose to act in ways other than the manner in which they have been conditioned, and that scientific and technological change produced social change and that this has become the most important human environment.

The articles in this volume deal with the ways in which science fiction has used (and misused) chemistry and the other sciences. But there is another side to the story that readers of this volume might consider as they think about the essays: that is, how science fiction has influenced science, particularly chemistry. Before you discard that idea out-of-hand, let me cite a few arguments. Way back in 1928 Jack Williamson, a pioneer author of SF who still is active (his novel, *Demon Moon* was published in 1994), won a contest with his article "Science Fiction—Searchlight of Science," based on the notion that science fiction in its speculations about the future illuminates possibilities that scientists can then follow up. I'm not sure Jack would want to defend that same statement today, but science fiction and science, particularly technology, have interrelated in surprising ways. In fact Frederik Pohl and I have struggled for years to put on a conference we called "The Shape of Things to Come" to bring leading scientists and SF writers together to discuss those interrelationships and how both groups can work toward a better future.

SF writers have always picked up ideas from scientists, particularly from those who are willing to speculate—and that includes a great many more today than it once did. As personal examples: I got my idea for *The Immortals* from scientific speculation about why cells age: for *The Listeners* from Walter Sullivan's *We Are Not Alone*, an account of attempts to pick up radio signals from the stars; and for *The Dreamers* from scientific accounts of chemical memory, discredited though they may be today. And every author of serious SF does the research necessary to provide a plausible scientific background for his narrative, and in so doing passes along to readers information about science, a belief in the importance of science, and the possibilities for drama and excitement in a scientific environment. However, scientists and inventors also have picked up ideas from science fiction. Consider:

- Konstantin Tsiolkovsky, the Russian rocket scientist, wrote, "The first seeds of the idea were sown by that great, fantastic author, Jules Verne; he directed my thought along certain channels, then came a desire, and after that, the work of the mind."

- Igor Sikorsky became interested in inventing the helicopter from reading Verne's *Robur the Conqueror*.

- Speleologist Norman Casteret said that Verne's *Journey to the Center of the Earth* first put into his head the idea of cave exploration.

- Lucius Beebe, Yuri Gargarin, Guglielmo Marconi, Santos Dumont, and many others said they had been inspired to their accomplishments by Verne's work.

- After a flight to the South Pole, Admiral Richard Byrd said, "It was Jules Verne who launched me on this trip."

- Submarine developer Simon Lake began his autobiography with the words, "Jules Verne was the director general of my life."

- Leo Szilard, who helped get the Manhattan Project authorized, said that it was H. G. Wells's *The World Set Free* that first started his thinking about the creation of an atomic bomb.

- Scientists such as Carl Sagan have testified that they were directed toward careers in science by their early reading of scence fiction, and I can remember a conversation with an anthropologist who told me that he chose that discipline because it was the closest thing he could find to science fiction.

- Scientists like my anthropologist colleague also have chosen their careers because of SF stories: A former director of research at the U.S. Bureau of Standards is on record as saying that it was the example of the fictitious Richard Bollinger Seaton (portrayed as a Bureau of Standards chemist in E. E. Smith's 1928 SF novel *The Skylark of Space*) that made him decide *that* was a life worth living. And when Fred Pohl spoke to several hundred undergraduates at M.I.T. and asked how many of them were first interested in science by reading science fiction, nearly every hand in the room went up.

- From its earliest days NASA was peppered with SF readers and John Campbell, longtime editor of *Astounding,* recalled that the circulation of his magazine in an obscure little town in Tennessee jumped from two or three copies each month to several hundred in the early 1940s, and it was only after the end of World War II that the name Oak Ridge meant anything uniquely important.

- Werner von Braun was so addicted to science fiction that he ordered a subscription to *Astounding* sent to a mail drop in neutral Sweden and forwarded to him at the research installation in Peenemünde. Evidence supports the speculation that for many scientists the discipline of science must first be preceded by the romance of science.

- Many inventions, from Buck Rogers's backpack rocket to robots, lasers, and computers, have first been described in SF stories. The discoverer of royal jelly, the food that turns female bees into queen bees, credited his inspiration to a speculation in a story by David H. Keller. Hugo

Gernsback, founder of the first SF magazine, *Amazing Stories*, predicted many developments, including radar, in his 1911 novel *Ralph 124C41+* and other writings, and he conducted a long (unsuccessful) campaign to permit the patenting of inventions described in SF stories, because, he said, conceiving them was ninety percent of the task of invention.

And Isaac Asimov once commented about the moon landing, "No one can say that science fiction writers and readers put a man on the moon all by themselves, but they created a climate of opinion in which the goal of putting a man on the moon became acceptable."

Maybe Jack Williamson's 1928 hyperbole was not so extreme after all .

Let me close with the hope that this is only the first of a series of books about the relationships between science and science fiction, and that Frederik Pohl and my "Shape of Things to Come" conference will someday come to pass.

Meanwhile, I would welcome any anecdotes, personal or general, about the way in which science and science fiction have influenced each other. Send them to:

JAMES GUNN
English Department
University of Kansas
Lawrence, KS 66045

An Overview

A Science Fiction Primer for the Uninitiated

Jack H. Stocker

Unquestioned Popularity

Science fiction (SF) is popular. If I say to you "Live long and prosper", the reference is immediately recognizable, and the offer of a shared enthusiasm extended.

Consider the weekly reports of the 10 best-selling titles of paperback fiction and understand that I'm using a very broad definition of science fiction. It is a rare week that does not include at least one SF offering and, on occasion, as many as four may be listed. A representative listing included two SF entries by a single author (Michael Crichton with *Sphere* and *Jurassic Park*) as well as other "SF-genre" entries.

Review the current movie and television offerings. Irrespective of quality, SF quantity is most amply provided. My city (New Orleans) has a round-the-clock SF channel. Think for a minute how much material is required just to fill that much broadcast time.

Consider the current spate of paperback-swapping bookstores. Almost all of these have strongly restrictive practices. If you want science fiction, you must offer science fiction, and no other category of book (mystery, western, espionage, romance, or biography, whether fiction or nonfiction) will be accepted for such exchanges. However, you may obtain any of the other categories in exchange for your SF books. This policy is based on hardheaded business practice: the recognition that science fiction sells well.

An advertisement in an issue of the *New Yorker* magazine offers an illustrated catalog, for $25, of a forthcoming auction of SF books. Obviously, this offer reflects an interested (and clearly affluent) audience. Twenty-five dollars is not yet "peanuts".

Be reminded how often SF characters and environments are used in making points. An effective and excellent example by the well-known editorialist Dick Locher has a Boris Yeltsin substitute for Captain Kirk on the *Enterprise* bridge, suggesting warp speed ahead for the new Russian democracy while a Spock-eared figure points out that this is illogical; the megabucks fuel is lacking. The winter 1994 issue of *Visions*, a journal issued by the Oregon Graduate Institute of Science and Technology, was largely dedicated to exploring the new information highway. Its front cover was boldly emblazoned with the headline "Beam Me Over, Scotty." Almost everyone will recognize the reference in both cases.

And perhaps most enlightening but not to be examined too closely is a report by the head of the New Orleans Public Library's information and reference division to the effect that best sellers do not have a very high theft rate, whereas Civil Service exam books, science fiction, and Bibles are the most likely to be stolen. Regardless of the reprehensibility of such actions, they do support the contention that science fiction is popular.

Dubious Respectability

Such enthusiasm, notwithstanding, is not reflected in public approbation. To paraphrase the complaint of comedian Rodney Dangerfield, "Science fiction don't get no respect." The concern is effectively illustrated by two anecdotes.

On being asked if he read science fiction, a man replied that he did not. When queried further as to whether he had read *Brave New World* or *1984*, he said of course he had. On having it pointed out to him that these were science fiction, he protested that they were not science fiction, they were *literature*.

A highly respected SF author, the late Theodore Sturgeon, was asked by a friend why he, Sturgeon, read that science fiction junk. After all, "90%

of it is crap." Sturgeon's famous reply, widely quoted and known as Sturgeon's dictum, was that "90% of *everything* is crap!"

Several major SF writers (including most notably Kurt Vonnegut) have militantly protested their inclusion in the "science fiction ghetto". Probably no area is more rife with noms de plume, often considered a way to protect one's identity from what might be considered embarrassing actions. One major SF writer has been identified with more than 30 such disguises.

And yet, on the other hand, a surprisingly large number of mainstream writers have at some point in their career provided us with a science fiction or fantasy story. *The Magazine of Fantasy and Science Fiction* has routinely published these minor efforts from major authors for years.

Elusive Definition

So, what is science fiction? Can we provide a clear statement of the territory claimed for the genre of science fiction? Unfortunately, no; attempts to do so invariably prove frustrating and unsuccessful. Certainly we can point to some "obvious" examples, *Star Wars* and *Star Trek* being perhaps the most widely recognized. Do these two series conform to the most frequently offered definition of "science fiction/Sci-Fi/STF/SF" as fiction based on logical projections from currently established scientific principles? Have we any basis for the faster-than-light (FTL to the SF buff) travel in *Star Wars*? Have we any scientific underpinning for the time travelers in *Star Trek*? Obviously we do not, and these and other cherished assumptions that have absolutely no scientific validity play a major role in many SF stories.

Is there a clear borderline between science fiction and fantasy? Again the situation is untidy. Can "magic" operate under rigorous rules? Certainly! Arthur C. Clarke, the author of the story on which the magnificent movie *2001: A Space Odyssey* was based, is credited with the saying that "sufficiently advanced technology is indistinguishable from magic". Often the distinction is merely a matter of whether the phenomena are "explained" regardless of the degree of double-talk involved. If one "explains" vampirism, a favorite fantasy/horror subject, as an illness involving a symbiosis between an invading organism and the human host, with the traditional cringing away from a Christian cross as due to the afflicted's residual superego response, have we not transferred the story out of the fantasy category? How did the author establish the ineffectiveness of a Crucifix against some of the afflicted? It turns out they were Jewish and could be protected by a visible Star of David. The story, by Richard Matheson, originally entitled "I Am Legend", was brought to the screen in an interesting but totally unrecognizable version scripted by John and Joyce Corrigan, the latter a Ph.D. physical chemist.

A sardonic quote from the distinguished SF writer Brian Aldiss (1) effectively encapsulates the problem.

> Anything that can be regarded as marvelous or as scientifically likely (or, on the other hand, as scientifically unlikely) has been press-ganged to serve under that amorphous term *science fiction*.

An alternate phrase, translating SF as "speculative fiction", has had wide circulation to circumvent some of these problems and has significant merit. This concept solves some but not all of the definition problems. It does appreciably broaden the definition and hence the coverage. It has been pointed out that the problem is not one of a definition of a single SF genre, but rather that there are a number of SF sub-genres, each of which, in turn, permits a reasonably coherent definition.

An agreeable summation of the problem was provided by a Swedish fan, Sam Lundwall, in his useful book *Science Fiction: What It's All About*, (2) as follows:

> The melancholy fact is that there does not exist any unitary definition of the genre. Or rather, there exist about as many perfectly valid definitions as there are readers of what I here for simplicity's sake call science fiction. (For myself, I would prefer the term Speculative Fiction as being more descriptive.) The SF buffs present in this connection certain resemblances to a select club where the venerable old men in the reading room have sat and slept in their moldering easy-chairs since the early twenties, with *Amazing Stories* and *Astounding Science Fiction* over their white heads; this is the Old Guard, which reads their science fiction with the emphasis on *science*, expecting nothing in the way of purely literary merits and, consequently, getting nothing of that kind. Every deviation from the rule of scientific accuracy is a scathing sin against all decency.
>
> The lovers of Space Opera are huddled behind enormous piles of *Startling Stories, Captain Future Magazine, Thrilling Wonder Stories*, and the collected works of E. E. Smith, and follow with glowing eyes the latest super-scientific adventures of the glorious Space Patrol in the Crab Nebula, where green BEMs [bug-eyed monsters] of the most atrocious sort are plotting

vile schemes against Humanity. Atomic Blasters blast, heroines cry, and the space ships leap in and out of hyperspace like frightened hens.

Right by, one can discern the Horror lovers with their blood-curdling *Weird Tales* and H. P. Lovecraft. European members of this group might be more fond of E. T. A. Hoffmann. They are a small and persecuted minority, far from loved by the *Amazing* readers.

The Fantasy and Sword & Sorcery groups are crowded together in a small room behind the reading room, from which they look rancorously out toward the sleeping gentlemen, thoughtfully fingering at their gleaming broadswords. They are also a minority, but literally acceptable since the recent upswinging interest in adult fantasy, and in strong need of *lebens-raum*.

The group of social reformers sit by the bar, where they exchange views on the future overpopulation, the food crisis, environment pollution, the goal of humanity etc., anxiously watched by the H. G. Wells phalanx, which stands somewhere between the reading room and the bar and doesn't know exactly where they belong.

The "New Wave" advocates keep themselves company out in the cloakroom. This is a collection of bearded and long-haired persons who experiment with new literary forms; they are loud and bothersome and do not have deference for anything, not even for the founder of the club, old Uncle Hugo Gernsback, and they are regarded with deep distrust by all other members. Some of them are said to be supported financially by the Establishment. The members of the science fiction community are deeply worried.

And yet all those factions and branches are only different sides of the same coin.

So do we include the horror cultists, the swords-and-sorcery buffs, the space opera enthusiasts, the utopia/dystopia reformers in addition to the "hard science" purists? One may choose whatever territorial milieu most pleases himself, herself, or itself, but soon recognizes that the choice is not a consensual one. It becomes simpler and certainly less frustrating to consider, as did the SF historian quoted, that all these various SF citizens simply occupy different rooms in the same house, pursuing their narrow interests, uninterested in or unaware of their neighbors. Perhaps a brief paragraph or two directed to each of these subdivisions might prove useful.

Hard science fiction occupies the top of the pecking order and invokes the names of Jules Verne and H. G. Wells, succeeded in modern times by Isaac Asimov, Poul Anderson, Arthur C. Clarke, Robert Forward, Robert Heinlein, and Larry Niven, among many others. The stories are science-driven even when the scientific basis is shaky. The category permits a number of further subdivisions, often reflecting the exploitation of some newer development. For example, in the late 1940s following World War II and the advent of atomic radiation, the consequences of mutation became (and remain) a popular theme for SF authors to explore. Two newer areas receiving significant current attention can be identified:

- Cyberspace/virtual reality, with particular attention paid to the grittier personal consequences, giving rise to the "cyberpunk" identification. William Gibson is its most identifiable guru; the movie *Johnny Mnemonic* is adapted from his novelette of that name.

- Nanotechnology, the designing of instrumentalities that operate on the level of atomic sizes and distances and the consequences thereof.

These two newer subdivisions are not mutually exclusive and often coexist within the same story. They can be considered as subsets of High Tech. For further guidance, see section "Recommended Reading".

Social science fiction is one of the largest categories and has a long and highly respectable history. It is the territory of utopias or (more frequently) dystopias, and was often written to promote religious or political ends. Widely known examples include Ralph Bellamy's *Looking Backward*, H. G. Wells's *Men Like Gods*, Aldous Huxley's *Brave New World*, and George Orwell's *1984*.

Space opera is named for its obvious correlation with "horse opera", differing from the latter's posse riding around a mesa rock, six-guns blazing, only in that the spaceship is blasting around an asteroid, ray-guns blazing. The category is widely beloved and can have a sophistication appropriate to all ages and backgrounds. The *Star Wars* trilogy is clearly a prime example.

Swords and sorcery has its damsels, demons, knights, and barbarians. Obviously, it includes Conan and Tarzan but is not limited to them. Much the largest portion of "S and S" is outright fantasy, but it has been melded very successfully with hard science fiction by a number of authors. At its broadest definition, it provides a very major contribution to the field.

The general category of **psi phenomena** deals with extrasensory perception (esp), telekinesis, teleportation, clairvoyance (which has been suggested to be telepathy through time!), pyrokinesis, and a large number of other abilities sometimes referred to as "wild talents". Again, the classification of a story as science fiction or fantasy may really rest on the author's choice of explaining the phenomenon or merely reporting it.

It is convenient to lump the "what might have been" with the "alternate histories" and "branching time-stream" stories. The questions posed by history buffs of "what if" (the Confederacy had won the War Between the States, Napoleon had been assassinated, or Lincoln had not) have produced some quite splendid writing in which trends, evidenced at the time of the historical nexus, in contrast to our history, withered or flowered. A number of very prominent writers (e.g., Winston Churchill) have indulged themselves in this fashion (3) and at least four collections of such stories are currently available under the overall title *What Might Have Been* (4).

The **horror** category is most conveniently considered to cover those stories dealing with the supernatural. Such a definition would permit inclusion of many of the primitive or religious myths that involved supernatural beings, for example, the ancient gods and their dealings with humans. For matters of convenience, Mary Shelley's *Frankenstein* and Bram Stoker's *Dracula* are usually considered the category's most recognizable ancestors. Stephen King is certainly the all-time most successful practitioner in this area (his book *Danse Macabre* is a masterly analysis of the field, dealing with both his own writing and that of others; it can be highly recommended for its perceptiveness and lucidity), but the shadow of Edgar Allan Poe and, even more, that of his lineal descendant, H. P. Lovecraft, are still very much with us.

The apparent distinction between "supernatural" and "scientific" has become a challenge for many writers. When one "explains" an apparently "supernatural" phenomenon in "scientific" terms, does the story involved then become science fiction? Various authors have had their protagonist challenge (and defeat) "supernatural" forces by their intellectual skills and scientific knowledge. Can a demon survive a direct hit by a 10-kiloton A-bomb?

The preceding material should effectively underscore the futility of a generally satisfactory definition of science fiction. Various name changes and hybrid nomenclature have been proposed to solve the problem, yielding, for example, "science fantasy", as well as the previously mentioned "speculative fiction", employed with only limited success. It becomes simplest to consider either the most generous definition or, to paraphrase Lewis Carroll, and say "Science fiction is what I say it is."

(A small sidelight: In its earliest days, during the 1920s and 1930s, science fiction was abbreviated as "STF", short for scientifiction. To the knowledgeable it became (and remains) SF; the term "Sci-Fi" is an abomination to SF enthusiasts. The advent of the Sci-Fi cable TV channel has probably effectively destroyed any possibility of slaying that particular monster. Those who feel strongly enough about it may employ the hard "c" in "sci" and call it the "skiffy" channel.)

Indeterminate Beginnings and Subsequent Meanderings

In view of this broad definition, the beginnings of science fiction also become accordingly imprecise. Did it begin with the myths handed down by the Greeks and Romans and their predecessors? Do we include the Bible? (e.g., was Ezekiel's heavenly fire the rocket trails of extraterrestrial visitors?) How about the monsters in medieval plays? Were they perhaps also from other planets?

For such reasons, excluding only the works of a very few earlier authors, it is convenient, if arbitrary, to start any discussion of SF with the late 1920s in the pulp magazines. Edgar Rice Burroughs, with his tales of Tarzan and John Carter of Mars, was very much with us as well as a number of his contemporaries writing of their protagonists stumbling into assorted lost civilizations in this world and elsewhere. Stories that we would today classify as SF/fantasy appeared sporadically and essentially uniquely in a number of magazines, almost invariably their only appearance. None of these publications were dedicated solely to this type of story. Beginning in the late 1920s, three uniquely science fiction magazines appeared:

- *Amazing Stories*

- *Astounding Stories* (metamorphosed through ASTOUNDING *Science Fiction, into Astounding SCIENCE FICTION* and finally to *Analog Science-Fact, Science Fiction*, all abbreviated comfortably as *ASF*)

- *Wonder Stories* (changed to the more melodramatic *Thrilling Wonder Stories*)

These three magazines dominated the field for the next 20 years. (Two of them are still with us!). Perhaps the most significant event of that time was the selection of John W. Campbell, a major science fiction writer in his own right, as editor of ASF in 1937, a position he held for 34 years. Campbell ushered in the modern age of SF with his insistence that in addition to involving innovative, thoughtful themes, ASF stories had to be at least minimally literate. He is credited with developing an impressively large number of the authors who defined the field for the next several decades: Isaac Asimov, Robert Heinlein, and A. E. van Vogt, among others.

A detailed history of the SF field for the past (approximately) 50 years offers some fascinating reading, which space limitations preclude detailing here. (Recommended further reading is offered in the chapter of that name.) Here, however, I offer some brief supplementary ramblings.

The 1940s saw the emergence of the problem-solving stories. What do you do if you get locked out of your spaceship (recall this problem in the movie *2001: A Space Odyssey*), or if your fuel is suddenly inadequate (employ a slingshot effect, exploiting some terrestrial body's gravity)? The popularity of this kind of story is still very much with us and is a recurrent *Star Trek* theme.

The 1950s saw a liberation from the limited outlets accepting SF material. There was a proliferation of new pulp magazines, two of which were to play very significant roles: *Galaxy Science Fiction* and *The Magazine of Fantasy and Science Fiction*. The latter, still very much with us, made no attempt to be purist as to the category in which its contents should be classified. The paperback field blossomed during this decade and PBOs (paperback originals), stories that had not previously appeared in the pulps, became increasingly common, although their hard-cover editions still were not of major importance.

The 1960s saw the advent of the "new wave", a movement that insisted there should be no restrictions on themes, writing styles, and so on. Previously, most SF writing, even at its most melodramatic, could be characterized as traditional—having a theme, employing complete sentences, and so on. Science fiction readers have tended to be a highly conservative group, many of whom found the new wave puzzling, often incomprehensible. But the dam had been broken, and the traditional "science fiction ghetto" has not been the same since. Much has been written implying there are "requirements" for successful science fiction writing, *imposed by the readers*, and the SF neighborhood is a kind of ghetto in which the inhabitants reside by choice. These readers consider the writers as their property and do not tolerate deviations from these unwritten rules. This phenomenon has provoked several prominent authors, e.g., Kurt Vonnegut, Harlan Ellison, and Bill Pronzini, to protest militantly that they are not science fiction authors and to insist that their books be regarded as mainstream efforts.

The 1970s and 1980s have seen the ascendancy of fantasy writing, to the point that it overwhelmingly dominates the field, a development not widely admired in the face of so little of the offered material being innovative.

The 1990s? The controversial Harlan Ellison (recall the gut-wrenching movie made from his even rawer *A Boy and His Dog*) has pointed out, in his favorite confrontational style, that the SF/fantasy field is simply publishing too much junk, and the publishers' belated recognition of this fact is about to lead to a drastic curtailment of the number of titles offered annually.

The problem is further exacerbated by the greatly increased offering of series, particularly the type in which the same characters interact in a number of stories. Each novel is self-contained, but the full story's denouement awaits the final volume. Trilogies abound; a "dekalogy" can be noted (*Mission Earth* by L. Ron Hubbard of Dianetics and Scientology notoriety); and there is a 30-volume series authored by an able English writer, E. C. Tubb, in which his character Dumarest desperately tries to find his way back from the stars to a lost and forgotten Earth. Such series might be regarded as the natural evolution of the classic serials without the melodramatic stops along the way. There is, of course, the more traditional type of series that customarily involves either the ongoing adventures of an individual or group or the establishment of an ongoing environment against which separate episodes can be played. Such series can build into an impressive totality; notable among them are the Darkover series by Marion Zimmer Bradley, the Pern series by Anne McCaffrey, the Childe Cycle series by Gordon Dickson, and, of course, the future history series discussed in the section "Authors and Themes".

To further appall the long-time SF reader is the current swamping of the SF field with spin-offs and copycat publishing. More than 150 *Star Trek* novels are derived from the original programs alone, and new ones are still appearing. The more recent three *Star Trek* series (*Star Trek: The Next Generation* or STNG, *Star Trek: Deep Space Nine* or DS9, and *Star Trek: Voyager*) each have burgeoning offerings of supplementary novels, concordances, starship blueprints, and almost any "et cetera" you can think of. The copycat problem is also effectively illustrated by the current spate of vampire stories spawned by the success of Anne Rice's efforts in that area.

A sidelight of interest: the SF field is possibly unique in that almost all of the successful older material is still in print. It's a rare novel by a major SF author that stays out of print for any length of time. New editions, for example, of the multiple-volume Skylark/Lensman series (written by "E. E. Smith, Ph.D.", a chemist), which appeared in the pulps from 1928 to 1948, continue to appear, often using the same plates and differing from the previous edition only in the cover art and, of course, the increased cover price.

This series, with its mind-boggling concepts and cardboard figures, is considered to have initiated and defined the category of space opera.

Authors and Themes

Science fiction writers are named, literally, from A to Z; we have Aldiss, Anderson, Asimov, and Anthony, on to Wolfe, Wyndham, and Zelazny. The territory they cover is even broader, but some of them have attempted to contain in time and space some of their own writing by using it to delineate a "future history" in which each of the stories (short story, novelette, novella, or novel) illuminates one segment. The stories rarely appear in chronological sequence, yet they permit an author to reuse an environment he may have painstakingly created while requiring him to be totally faithful to it. Robert Heinlein, Larry Niven, Gordon Dickson, Isaac Asimov, and Poul Anderson have been notably successful in these endeavors. Some authors have distinguished themselves by extensively exploring single themes; Isaac Asimov on robots or on overpopulation is a good example.

Choices of Media

Science fiction is media-diverse. It is available as the printed word in magazines and soft- or hard-cover books. It can be heard on audio cassettes and seen and heard as cinema and television (both types available on cassettes for control of viewing).

Magazine presentation is no longer limited to prose. The comics (traditional, underground, and all varieties between) appear in an incredible array of choices, some with print orders of hundreds of thousands. They have spawned entire conventions dedicated solely to the comics category and have created collectibles that, in "newsstand mint" condition, command prices up to and including $100,000. (So much for the Superman comics your mother threw out!) If one classifies the superheroes category as SF, then SF dominates the comics field and would quickly overburden a full chapter dedicated to exploring it.

In more recent time, recognition of a category, "science fiction art", has become firmly established. At one time, cover artists of paperback books were not even identified. This situation is no longer true; it is now recognized that the choice of cover artists can improve the sale of a book, and some collectors are interested only in the covers of SF books. Several quite splendid book collections of the covers of SF magazines exist, as well as single-artist collections of cover and internal artwork done by some of the more distinguished practitioners. The artists' offerings can range from the sublime (e.g., a Chesley Bonstell painting presenting in great detail, complete with appropriate light and shadow, the planet Jupiter as seen from one of its moons), to the grotesque (both SF and fantasy), to the whimsical, examples of which would include the Christmas covers of *Galaxy Science Fiction* magazine, customarily featuring an alien Santa Claus, perhaps with four arms, in his deer-shaped space ship, delivering presents to nonhuman children. A few recommended examples are listed in the Appendix.

A modest but recognizable category of SF-oriented music is familiarly known as "space music", occasionally written by major composers. It ranges from the soaring (and borrowed) music of the movie *2001: A Space Odyssey* to the simply familiar, reflecting the movie and TV themes of favorite offerings. A current TV ad offers a multiple cassette collection of *Star Trek*, original and new-generation, music, probably including Uhura singing "Somewhere Beyond Artares" and Spock playing his Vulcan harplike instrument. A further subcategory, "filksinging", is discussed in the section Science Fiction Fans), and an out-of-this-world cookbook (5) has been published, presumably dedicated to the sense of taste. We still have no SF entries catering to the sense of smell or touch, although both have been widely treated in prose.

In recent years, the science fiction domain of gaming (that is, SF- or fantasy-based role-playing games and board games) has mushroomed; it is now a significant component of the SF territory. Entire conventions are devoted solely to gaming activities. "Dungeons and Dragons" ("D-and-D") initially dominated the field (which is still predominantly fantasy-oriented), but current gaming activity now offers a much broader choice and is one of SF's fastest growing areas. Currently fashionable is "live gaming", distinguished by active physical participation by the players as contrasted with the static, seated-at-table traditional involvement. It deserves more coverage than space permits here.

Science Fiction Fans

Science fiction fandom is probably the least recognized aspect of the SF world. It spans the entire age range from "sub-teenyboppers" (ages 10–12) to septuagenarians and includes language, predominantly acronymic, unique to it. Some examples:

- FIAWOL (pronounced fee-ah-wol), "fandom is a way of life", which may be contrasted with FIJAGH (fee-jag), "fandom is just a goddamn hobby".

- BEM, bug-eyed monster. In the late 1930s and early 1940s, a group protesting the more lurid covers founded the SFTPOBEMOCOSFM or the Society for the Prevention of Bug-Eyed Monsters on Covers of Science Fiction Magazines.

- GOH, guest of honor (at a science fiction convention). This has a number of variations such as FGOH, (pronounced fan-go, word-play not intended) for the fan guest of honor, reflecting acknowledgment of his or her fannish activities. The plural of fan, incidentally, is "fen".

- GAFIA (pronounced gaf-ee-yah), "getting away from it all." Originally this meant getting away from daily dreariness into the world of science fiction. It has now evolved into meaning exactly the opposite: getting away from your fellow SF fanatics back into a more rational world.

- SMOF, "secret masters of fandom," reflecting a widely held paranoid view that a small group of money-hungry businessmen manipulate and exploit SF fans.

- TANSTAAFL (tans-tah-ful) is taken from one of Robert Heinlein's stories and translates to "there ain't no such thing as a free lunch".

- FILKSINGING. A significant number of SF buffs enjoy the collective singing of risquè lyrics of a science fiction/fantasy nature, the rawer the better, to familiar tunes. This activity of singing "filthy folk songs" is called filksinging.

- SCA, the Society for Creative Anachronism, has a large number of chapters across the country, and the members are deeply involved in all aspects of historic chivalry, including period costuming and appropriate armor. They put on pageants, tableaus, and so on, and are a common component of SF conventions.

The widespread and driving need for fans to express themselves is vented through their fannish publications, known as "fanzines" and APAs (Amateur Press Associations). The subcategories are legion and mostly self-explanatory in their names; newszines, genzines (gen for general), prozines ("pro" for professional), clubzines, Trek letterzines, and myriad others. They are more often traded than sold. The better ones enjoy appreciable respect.

Obviously there are fan clubs, usually tailored to a particular commitment. Some examples include "The Companions of Dr. Who", "Event One", "Krewe of the Enterprise", "Ista Weyr" (one of many Weyrs across the country, dedicated to the "dragon" stories of Anne McCaffrey) and, of course, less exotic titles such as Star Trek fan clubs, Anne Rice fan clubs, and so on. The University of New Orleans Science Fiction and Fantasy Student Club, to which I have been faculty advisor for a number of years, initially entitled itself SOB^2 standing for "Sons of the Big Bang". Accused of male chauvinism, it eventually retranslated its name to "*Survivors* of the Big Bang".

A major goal of the more committed fan is to attend as many SF conventions as possible. Most such conventions, even the largest, are predominantly volunteer efforts, so the opportunity to be a part of the planning and implementing of such affairs is easily found. The conventions ("cons") range from half-a-day to week-long, with attendances ranging from less than 100 to more than 15,000, including an annual World Con, often held outside the United States. The cons are frequently given names involving word play. Some examples are

- Vul-Con (held in New Orleans for 14 consecutive years, with a *Star Trek* emphasis)

- Magicon (Orlando)

- I-Con

- Con-Stellation

- Chattacon (Chattanooga)

- Ecuminicon

- Among the most inventive, a series of conventions in Nashville: Kubla Khan Klave, Kubla Khan Too, Kubla Khan Kubed, Kubla Khan Kwandery, Kubla Khan Khanception, Kubla Khan Sex, and Kubla Khancensus.

A typical three-day SF con traditionally offers a costume contest; a dance (or two); a Dealers' Room; an extensive area devoted to gaming; a 24-hour movie room featuring new, old, and offbeat SF; one or more additional video rooms individually dedicated to showing videos of a cult-status TV series (e.g., *Dark Shadows*, *Blake's Seven*, or *The Invaders*); filksinging; and multiple panels, usually involving the convention guests, dedicated to both serious and frivolous subjects, how-to topics (designing an alien life form, writing for a particular market), readings from unpublished works, personal reminiscences, and so on.

It is not unusual to have three choices of presentations at any one time during the daytime hours. These guests constitute a major draw of the conventions, and there is competition to attract the most famous writers, artists, TV stars (they can command quite large fees for their appearances), TV and movie directors, and so on. The annual Dragon Con has scheduled more than 180 such Guests for its 1998 Atlanta Celebration. Even the more active fans are recognized on such occasions as Fan Guests of Honor. Quite frequently some function is held, most often an auction of donated items, the proceeds from which go to a deserving charity. Customarily an art show is presented, where amateur (and professional) art is displayed, much of it offered for sale in "silent auction" fashion. Other innovative planning is very common. Registration for three days of such a typical con escalates from $15–20 if paid well in advance, by stages, to $30 or more at the door.

Obviously, for a modest price, there is something for everybody.

In addition to conventions, a very large (and rapidly increasing) area of SF fandom activity involves the ubiquitous computer. Any relatively recent personal computer of modest capacity equipped with a modem links all interested parties (bereft of gender and age if they wish) into a number of networks of fellowship via the Internet. I was told several years ago that of the many hundred special interest groups titles available at least dozens were dedicated to SF interests. Belonging to some of them brings you enough electronic mail to overtax your printer and overfill your available time to read or process it. (Some such groups have made the plea that "trivial" communications be avoided because they go to every member affiliated with that special interest group). And remember! All of this is free to anyone accessing via a covered group, such as most university employees. Even for those not covered, the telephone charges are usually modest. Would you like a full printout of *Star Trek: The Next Generation* programs, in chronological order with the original air dates? Someone has done it as a labor of love and was willing to share it (for free) with his fellow enthusiasts in a particular interest group.

A recent edition of *The Internet Yellow Pages* (6) had some 11 entries under the Star Trek heading (in addition to those under science fiction and related classifications). One entry for Klingons offers instruction in the Klingon language and considers such topics as Klingon love poetry, haiku, and thoughts on Kronos as the homeworld.

The world of science fiction recognizes its most successful practitioners through a number of annual awards. Of the two most prestigious, the Hugo,

named after the respected pioneer SF writer and promoter, Hugo Gernsback, is the oldest. The awardees are chosen by those registered at a World
Con and may be classified in a very broad range of categories, for example,
Best Fanzine, SF Artist, TV Program, or Movie, in addition to the more
conventional awards for prose, based on story length. The winners are
clearly selected by the fans.

The other award, the Nebula, is limited to prose fiction and is divided
into novel, novella, novelette, and short story classifications. The winners
are selected by members of the Science Fiction Writers Association
(SFWA). (One of the contributors to this book, Connie Willis, carries the
distinction of being the only person to have been awarded Nebulas in all
four categories). A number of other, usually more specialized awards for
unique areas are presented, such as for fantasy or for first efforts.

In Summary

You should now consider yourself introduced to some of the arcane rites of
science fictiondom. The material should serve as a very brief overview of
the SF metropolis and its suburbs. One more bit of background is pertinent.
It consists of strongly underscoring the point that many of us were attracted
to the sciences by what is often called a "sense of wonder". The world was
clearly a wondrous place and SCIENCE (in capital letters) beckoned to us
with the promise of EXPLAINING EVERYTHING. We know what happened; in too many cases we were subjected in both high school and college
to a catechism that encouraged orthodoxy and formally discouraged speculation beyond the boundaries of the syllabus. During the later days of graduate research, we came to realize that our laboratory efforts dealt with a
minuscule part of one small territory in only one of the many divisions of
chemistry. Where could we get a sense of the larger, more meaningful picture? Not from reading *Annual Reviews*. From science fiction of course!

For those readers who already knew most of the foregoing, their indul-
gence of my messianic fervor is respectfully requested. For those who perhaps disagree with some of the more sweeping generalizations offered, I
enthusiastically invite and would welcome a dialogue. Those who might
find useful some specific guidance in exploiting the potential utility of science fiction in the classroom are invited to explore the appropriate subsequent sections of this book. And if, like me, you remember with nostalgic
pleasure the galvanizing impact of the earliest SF you encountered (books,

comics, movies, TV, or whatever), you are reminded that the attractions are still out there awaiting you.

References

1. Aldiss, Brian; in *SF Source Book*; Wingrove, David, Ed.; Van Nostrand Reinhold: New York, 1984, Foreword.
2. Lundwall, Samuel J. *Science Fiction: What It's All About*; ACE Books, New York, 1971, pp 15–16.
3. *If: or History Rewritten*; Squire, J. C., Ed.; Speculative Essays, 1931.
4. *What Might Have Been*; Benford, G., and Greenberg, M. H., Eds.; Bantam Books, New York, 1989. Combines Vol. 1: *Alternate Empires* and Vol. 2: *Alternate Heroes.* A second combined volume offers Vol. 3: *Alternate Wars* and Vol. 4: *Alternate Americas.* The series appears to continue with volumes entitled *Alternate Presidents, Alternate Kennedys, Alternate Warriors,* and *Alternate Outlaws,* all edited by Mike Resnick for TOR books. The novel *Bring the Jubilee*, describing an alternate America in which the South won the War between the States, is a particularly fine example. (Ward Moore, Avon Books, 1955). There are many others.
5. *Cooking Out of This World*; McCaffrey, Anne, Ed.; Ballantine Books, New York, 1973. (Recipes supplied by some well-known SF authors.)
6. Hahn, H.; Stout, R. *The Internet Yellow Pages*; Osborne McGraw-Hill: New York, 1994.

★2★

Science in Science Fiction

A Writer's Perspective

Connie Willis

★ If your only contact with science fiction has been TV and the movies, you may wonder how the word "science" ever got into "science fiction". In *Star Wars* loud explosions occur in outer space, and Han Solo brags that his spaceship did the "Kessel run in 12 parsecs". In *Total Recall* we find out to our surprise that Mars has a core of solid ice and that it takes less than a minute to give a planet a viable atmosphere. On *Star Trek: The Next Generation* we are supposed to believe that they have invented technology that can take them all over the galaxy, but they haven't figured out the seat belt yet.

Written science fiction shares the movies' reputation of far-fetched ideas and "made-up" science of ray guns, antigravity devices, and bug-eyed monsters, and sometimes it deserves it, but the word "science" in "science fiction" is more than just a euphemism for space opera.

Science is an integral part of science fiction, and science fiction writers are well-versed in science. As Hal Clement said (1) in his introduction to *First Flights to the Moon*, "Really, the only difference between science fiction and the rest of literature is that the former demands that its writers know more." Many science fiction writers are scientists themselves—Isaac Asimov was a biochemist, Arthur C. Clarke is a physicist and mathematician,

David Brin is an astrophysicist, and Philip Latham is an astronomer. Many others are noted for their careful research and attention to scientific detail.

The close connection between science and science fiction has been demonstrated over the years by a number of anthologies of stories written by scientists, such as *Science Fiction by Scientists*, edited by Groff Conklin. This connection has also been demonstrated by anthologies of stories about science, such as *Where Do We Go From Here?*, a collection of stories about science edited by Asimov; *First Flights to the Moon*, which contains comments about the accuracy of the science following each story, edited by Clement; and *Time Probe: The Sciences in Science Fiction*, edited by Clarke.

Most recently, Byron Preiss edited a series of four anthologies combining science and science fiction: *The Planets*, *The Universe*, *The Microverse*, and *The Ultimate Dinosaur*. These anthologies contain articles by eminent scientists such as Leon Lederman and David J. Helfand on topics ranging from quantum theory to the genetic code. Each essay is then followed by a science fiction story by a prominent science fiction author that incorporates the scientific material into imaginative speculative fiction.

The standards for accurate, up-to-date science are very high in all science fiction. Stories about cutting-edge discoveries and technologies must reflect the current knowledge in the field, and extrapolations must be based plausibly in present-day science. (A story such as Stanley Weinbaum's classic, "A Martian Odyssey", with its Earthlike atmosphere, canals, and birdlike Martians, would never be accepted by a science fiction editor today.)

Science is used in a variety of ways by science fiction authors: as subject, as plot, as setting or background, and as metaphor. It also serves as the underlying structure for a diverse, constantly experimenting, and expanding field.

Science as Subject

Science fiction is most well-known for using science as its subject. People frequently say that science fiction is "the literature of ideas", and some science fiction authors have gone so far as to define science fiction as fiction about science. Such a definition is very narrow and excludes many of the classic works of science fiction, but it certainly defines one type—the subgenre known as "hard" science fiction.

Hard science fiction uses science as its central focus and subject. Two examples are *The Andromeda Strain* by Michael Crichton and *Ringworld* by Larry Niven. *The Andromeda Strain* is about a virus from outer space that lands in a southwestern town and infects the inhabitants. The virus, and the scientists' attempts to understand it and control it, form the heart of the story. In Niven's *Ringworld*, a group of humans and aliens explore a giant ring that has been constructed around a star, but they and their adventures are secondary to the main character, the ringworld itself.

Other fine works of hard science fiction include Clarke's *Rendezvous with Rama*, about the discovery of a deserted spaceship; Kim Stanley Robinson's *Red Mars*, the first book in a complex series about the colonization of Mars; and *Neuromancer*, by William Gibson, which deals with computers and virtual reality. In these works, science serves as theme, subject, and plot. "Hard" science fiction is sometimes criticized for not having interesting characters, but I think that misses the point. The science itself is the main character, and it is a complex and fascinating one.

As George Turner (2) said, the speculative future situation set up by the author of science fiction is "both the nexus and ambience of his work; the [human] characters serve the purpose of demonstrating how the realities of his alternative will affect men and women."

The "literature of ideas" is nowhere more apparent than in the science fiction short story, which has a long tradition of exploring scientific concepts and their ramifications. Readers often refer to stories as "the one about …" or say, "There was this great story. I can't remember what it was called or who wrote it, but the idea in it was …". Among the classic science-as-subject short stories are Latham's "The Xi Effect", in which the universe begins to shrink while the laws of physics remain the same; and Asimov's "Nightfall", about a once-in-a-thousand years' view of the stars.

Some of these science-as-subject stories are only just barely stories. Asimov's "Pâté de Foie Gras" is actually a scientific explanation of the goose that laid the golden egg (*see* Chapter 13), and Niven's "Man of Steel, Woman of Kleenex", is a treatise on the physical effects of Superman's superpowers. The ultimate in this type of nonstory story is Asimov's hilarious "The Endochronic Properties of Resublimated Thiotimoline" (*see* Chapter 12), which is written as a learned scientific paper, complete with graphs, formulas, and footnotes.

Time travel and its inherent paradoxes are frequently the subject of science-centered stories. Mack Reynolds's "Compounded Interest" sets up an essential paradox: the closed-time loop in which the cause and the effect are one. In the story, a man travels into the Middle Ages to put a little money into a bank. The money accumulates interest over the centuries and becomes the fortune the man needs to build the time machine so he can go back to the Middle Ages to deposit the money.

Robert Heinlein's "All You Zombies" and "By His Bootstraps" and Charles Harness's "Child By Chronos" carry the paradox a step further by making child and parent the same person. Fredric Brown's "The Yehudi Principle" does even better than that. His story, which describes the test of a wish-fulfilling device, is itself a closed loop, brought about by a character's wish for a story that writes itself. It begins and ends at the same place.

Robert Silverberg's "Many Mansions" and David Gerrold's *The Man Who Folded Himself* have even more complications. The ultimate closed-time loop, however, was one proposed, but not written, by Stanislaw Lem. Lem (3) set up a loop using Richard Feynman's scientific theories in which the Big Bang is caused by a time traveler firing a positron back through time to the beginning of the universe.

Such stories are less stories than intellectual exercises. They are a kind of game in which ideas are played with, and it is the ideas that are interesting, not the human story. These stories have the same logic as an M. C. Escher painting, and the same fascination.

Science as Plot Device

Even when the story is not about science, science is frequently used as the basis for the plot. These stories are sometimes called "gadget" stories because the plot depends on the inventing or use of a scientific device. Examples are James Blish's "Beep", which uses positrons to send instantaneous signals, and Clarke's "A Slight Case of Sunstroke", in which parabolic mirrors are used to focus sunlight.

In John Stith's *Redshift Rendezvous*, a murder is committed on a spaceship traveling at near the speed of light, and the effects of relativity serve as clues to solving the mystery. Murder is the problem in Niven's "A Kind of Murder", and its solution depends on the laws governing heat and potential energy. "Special Delivery", by Carol Dieppe and Lee Wallingford, is another kind of mystery, with the solution encoded in the sequence of DNA.

The plot of H. Beam Piper's "Omnilingual" depends on the universality of the periodic chart. The plot of Brown's "The Waveries" depends on electromagnetic fields, and Heinlein's plot in *Time for the Stars* revolves around the theory of relativity.

In the 1930s, Ross Rocklynne wrote a whole series of stories with each plot literally a scientific problem, about a detective and the criminal he is chasing. In "The Men and the Mirror", criminal and detective find themselves helplessly sliding across the surface of a giant concave mirror and have to use their knowledge of pendulums to rescue themselves.

In some cases, science literally drives the plot. In Blish's "Nor Iron Bars", a spaceship is shrunk to the size of a subatomic particle and travels through the spaces of an atom as if the atom were a solar system. Asimov's *Fantastic Voyage* uses the same device, but this time the reduced vehicle travels through the bloodstream of the human body.

Science as Background

The most prevalent use of science fiction is as background or setting for the story. Examples of this are nearly every science fiction story ever written. The majority have, of necessity, a science-related background because of their settings—outer space and the future.

From Heinlein's *Red Planet* to Ray Bradbury's classic *The Martian Chronicles* to Robinson's ambitious new Mars trilogy, science fiction has set numerous stories on Mars and every other planet of our solar system. Niven's "Becalmed in Hell" is set on Venus, Poul Anderson's "Call Me Joe" on Jupiter, John Varley's *Titan* on a moon of Saturn, and Heinlein's *Have Space Suit, Will Travel* on Pluto.

Planets are not the only settings. Endless stories, from H. G. Wells's *The First Men in the Moon*, written in 1901, to "Griffin's Egg", written by Michael Swanwick and published in 1994, are set on the moon. Asimov's "Marooned on Vesta" takes place on an asteroid, and Bradbury's "The Golden Apples of the Sun" is set on a spaceship diving straight into the sun's corona.

Almost any place in the universe can serve as a setting, from a black hole, to the Lesser Magellanic Cloud, to space itself. Asimov's "Nightfall" is set in the middle of a globular cluster; Brian Aldiss's "The Dark Soul of Night" at the edge of a black hole; and my story, "The Sidon in the Mirror", on the surface of a cooling red giant.

Science permeates the stories. In Heinlein's *Have Space Suit, Will Travel*, Pee Wee's space suit starts to run out of air, and Kip has to solve the problem of air pressure and necessary oxygen levels. In Tom Godwin's "The Cold Equations", the pilot has to deal with the inexorable realities of too much mass and not enough fuel. In "The Sidon in the Mirror", the characters have to deal with high levels of explosive hydrogen in the atmosphere.

These stories are, of course, set in the future, as are many other science fiction stories that date all the way back to Wells's *The Shape of Things To Come*. There are as many futures as there are stories.

It is from these science-in-the-future stories that science fiction got its reputation for prediction, extrapolating everything from radar (Murray Leinster's "Politics") and the atomic bomb (Cleve Cartmill's "Deadline") to television (Hugo Gernsback's *Ralph 124C41+*) and the credit card (Heinlein's *Tunnel in the Sky*).

The focus, though, is on extrapolation rather than prediction: what might happen instead of what will. Science fiction is interested in the secondary effects of science as well as in the technology itself. Heinlein's *Waldo* explains an imagined future of radiant power, underground buildings, and remote controlled tools in great detail. R. C. Fitzpatrick's "The Circuit Riders", in which an emotion-reading circuit board aids the police in stopping crimes before they happen, only hints at the science and technology behind the device.

Science as Metaphor

Science is also used in science fiction as a metaphor by giving concrete form to intangible ideas and emotions. A good example is Bob Shaw's "The Light of Other Days", which extrapolates an invention called "slowglass" that uses the laws of physics to slow down the passage of light through glass. It takes light several years to traverse the width of a pane of slowglass; thus, allowing people to view scenes from the past. The story deals with the interesting ramifications of such a technology, such as windows set out in spring to absorb scenes of blossoming trees for winter sale to city dwellers. On another level, though, the story is about loss and grief and the persistence of memories of the past in our minds.

The genre is uniquely suited to examine ideas and emotions through metaphor. As Damon Knight said (4) in *In Search of Wonder*, "In science fiction we approach that mystery which surrounds us, not in small, everyday symbols, but in the big ones of time and space." Stories about time travel inevitably bring up issues of memory and regret and the desire to undo the past. Artificial humans automatically raise the issues of defining real humans and of the nature of reality.

Science fiction has the added gift of being able to create worlds or societies that give actual shape to emotional and philosophical issues. As Philip K. Dick said (5), science fiction lets a writer "transfer what usually is an internal problem into an external environment; he projects it in the form of a society, a planet, with everyone stuck, so to speak, in what formerly was one unique brain."

Kurt Vonnegut does this in "Harrison Bergeron", a story in which the concepts of equality and "fairness" in society have become reality. Ballerinas are forced to wear lead weights to make them equal to everyone else, and geniuses are fitted with helmets that interrupt their thoughts with electric shocks. The story gives an abstract concept a physical form and a powerful imagery that force the reader to reexamine the idea.

In a similar manner, Bradbury gives concrete form to the impulse to censor in *Fahrenheit 451*, the story of a "fireman" whose job is to burn libraries. Aldous Huxley's *After Many a Summer Dies the Swan* has an immortality drug that embodies our desire to live forever and shows us the emotional results of that desire in physical form. Dick's *Do Androids Dream of Electric Sheep?* and Asimov's "The Bicentennial Man" use androids and robots, respectively, to deal with the question of what is human.

The science in such stories is not just a hook on which to hang the symbolism; it is frequently there as subject and plot as well. Blish's "Surface Tension", the story of a group of scientists who are marooned on a harsh planet and whose only hope of survival is to transform themselves into microscopic aquatic creatures, is a scientific look at life in a tidal pool. The story also makes a serious statement about the tiny tidal pool we live in and the fascinating world beyond, and also about our own limited view of the universe.

Ironically, "Surface Tension" is frequently cited as a classic story of the "Golden Age", when science fiction was straightforward and without subtext or symbolism. The contention is that the use of science as metaphor did not enter the field until the 1960s, with the onslaught of the literary New Wave, but this stand is difficult to support.

Early writers of science fiction, particularly Wells, used stories of the future to illustrate present societal ills and that tradition has continued to the present day in such diverse works as Harry Harrison's "Make Room, Make Room!", which deals with the problem of overpopulation; Theodore Sturgeon's "And Now The News …", a chilling cautionary tale about the dangers of media saturation; and Gordon Dickson's "Computers Don't Argue", a humorous story about the deadly drawbacks of computerization.

Peter Nicholls, in his essay "Metaphysics", stated (6) that "the exploration of metaphysical questions has been central to [science fiction] at least since the time of Mary Shelley's *Frankenstein*", and that philosophical themes have always been as much a part of the field as scientific ones. *Frankenstein*, considered by many to be the first science fiction novel, was also the first to use the technique common in science fiction of giving concrete shape to an intangible idea. Frankenstein and his monster are a literal embodiment of science and man's inability to control his own technological discoveries.

The use of science as metaphor did increase significantly in the 1960s. Such authors as J. G. Ballard, Harlan Ellison, and Samuel Delaney used science fiction extensively to explore psychological and emotional realities.

In recent years a new form of science fiction story has emerged, in which the symbolism is reversed and the story functions as a symbol for the science. In these works, the surface story is ostensibly about characters doing sometimes quite ordinary actions, but these surface actions parallel a scientific principle or reality. This parallel is implied rather than stated outright, and the whole story serves as an extended metaphor for the science underneath.

Pamela Zoline's story "The Heat Death of the Universe", was probably the first story of this type. Written in an experimental style, it appears on first reading to be a listing of a housewife's life as she slowly falls apart emotionally. On another level, however, it is a detailed representation of the workings of entropy.

In 1977, Ed Bryant experimented with the same story structure in "Particle Theory". The story is ostensibly about an astronomer who has prostate cancer and is being treated for it with radiation, and reads very much like a mainstream story, but it is also the story of supernovas and a decaying galaxy. As the story progresses, the parallels between the astronomer's fate and the universe become increasingly apparent.

Bryant's story is about relativity theory and subatomic physics, and the choice of scientific concepts is no accident. This kind of story seems to be peculiarly suited to the complexities and uncertainties of these concepts. George Alec Effinger's "Schrödinger's Kitten" uses an extended metaphor to look at the implications of the Schrödinger's Cat thought experiment, and is itself a thought experiment. In my story "At The Rialto", I use the activities of a group of physicists at a conference in Hollywood to illustrate quantum theory. The physicists are looking for a paradigm for quantum theory, and they find it in such odd things as bimbos, the footprints at Grauman's Chinese Theatre, and the inability of a group of physicists to fix an overhead projector.

Other stories that use this technique include Eileen Gunn's "Stable Strategies for Middle Management", which parallels corporate politics and the insect world; Raccoona Sheldon's "The Screwfly Solution", which explains the violence between men and women in terms of pest-extermination technology; and my short story "Schwarzschild Radius", which compares black holes to their discoverer's experiences on the Russian front in World War I.

These stories not only illustrate science through the metaphor of human events, but show the connections between them. At their best, they can illuminate them.

Science as the Basis of All Science Fiction

The preceding sections present a few examples of the ways in which science is used in science fiction: as subject or plot, as background, and as metaphor. But in another way, every science fiction story has its basis in science.

Reginald Bretnor, attempting to define science fiction, described (7) it as a type of literature that is informed by a scientific awareness and a scientific approach and is based in a background of scientific knowledge. This definition has been called "the science fiction method", which Frederik Pohl described (8) as "a way of looking at any subject, taking it apart into its components, and putting it back together with some of the existing parts replaced by new inventions."

Judith Merril went even further. Speculative fiction, she said (9), "makes use of the traditional 'scientific method' (observation, hypothesis, experimentation) to examine some postulated approximation of reality, by introducing a given set of changes—imaginary or inventive—into the common background of 'known facts'." And Robert Conquest noted (10) that science fiction "ranges over every type of story in which the centre of attention is on the results of a possible, though not actual, change in the conditions of life."

Taken together, these definitions seem to capture the real truth about science fiction: that it is one huge thought experiment, with each author observing the world, developing hypotheses about it, and setting up experiments in the form of stories to examine those hypotheses. In this sense, every science fiction story uses science, whether science appears overtly in it or not.

The field is too large and complex to show the workings of this vast thought experiment, but I can give a couple of examples. The first is Asimov's Three Laws of Robotics, which he first introduced in the story "Runaround". These laws, which parallel Newton's Three Laws of Motion, were built into the hardwiring of the robots and govern robot behavior, making it impossible for the robot to hurt the human beings who built it.

These laws became the basis of not only Asimov's robot series—*I, Robot* and *The Rest of the Robots*—but of other writers' robot stories as they explored the ramifications, complications, and paradoxes of the three laws. In "Liar", Asimov shows us a telepathic robot who, because of the First Law, cannot injure human beings. Because it can read minds, it knows what the humans want to hear and is aware of the emotional hurt it would inflict by telling them the truth, setting up an apparently insoluble conflict in the robot.

In "Plato's Cave" by Anderson and "Blot" by Clement, the difficulties of obeying the First Law are explored when the concept of "human" is unclearly defined. Russ Markham ("The Third Law") and Lester del Rey ("A Code for Sam") have also written stories based on the problems of the Three Laws. My story, "Dilemma", has the robots going to Asimov and asking him to repeal the laws. Harrison in "The Fourth Law of Robotics", added another law. All of these stories set up experiments that test the Three Laws against the laws of nature and human nature in much the same way that scientists test Newton's laws: by introducing variables that either change the situation or look at it from a different angle.

The second example of using observation, hypothesis, and experiment is a broader one: writing stories about recurrent themes in the field. These themes include the android, time travel, the end of the world, and the coming of aliens to Earth. These themes have been with science fiction since its beginning, but instead of being played-out ideas, they are constantly rewritten and reinvented as writers "ring the changes" on them by introducing new variables and new angles of vision.

The theme most frequently identified with science fiction—alien invasion—has been written about in hundreds of stories and in hundreds of different ways. The classic approach, of course, is that of the aliens who want to kill all of us and take over our planet. It was first (and probably best) written about in Wells's *War of the Worlds*, and gave rise to dozens of "bug-eyed monster" stories and the classic "Who Goes There?" by John W. Campbell.

In Heinlein's *The Puppet Masters*, the aliens are parasites who want to use us as hosts. In C. M. Kornbluth's "The Silly Season", they want to enslave us. In Jack Finney's *The Body Snatchers*, they want to take over our bodies. Sometimes the aliens are friendly, as in Clarke's *Childhood's End*. Sometimes, as in Knight's "To Serve Man", they *pretend* to be friendly but are not.

Other times we are merely in the way. In Raccoona Sheldon's "The Screwfly Solution", the aliens view us as nuisances and set about exterminating us by using the same methods we use on bothersome insects. In John Wyndham's *The Kraken Wakes* and Clark Ashton Smith's "The Metamorphosis of Earth", they are not even aware of us as they blithely begin changing the planet to suit their own specifications.

In Mildred Clingerman's "Birds Can't Count", they want to observe us; in Margaret St. Clair's "Prott", they want to tell us their problems; and in William Tenn's "The Liberation of Earth", they want to use us as a battlefield in a war with somebody else. In Knight's "The Big Pat Boom", they want to buy souvenirs; and in J. G. Ballard's "The Watchtowers", they do not seem to want anything. They just sit there in the sky—silent, unmoving, inscrutable.

The list goes on and on. Some of the stories have a great deal of overt science; others none at all, but they are all informed by science. Each author performs a different experiment by looking at the problem from a different angle and introducing a different variable.

These individual experiments make up part of the giant, sprawling thought experiment called science fiction. Science forms its basis, its ethic, and its technique. It functions as subject, as plot device, and as background. It also functions as a metaphor for philosophical and psychological concepts, and in turn finds metaphors for itself. Science is thus central to science fiction, its subject and its setting, its source and its structure.

References

1. Clement, Hal. "Introduction". In *First Flights to the Moon*; Clement, Hal, Ed.; Doubleday: Garden City, NY, 1970.
2. Turner, George. "Science Fiction as Literature". In *The Visual Encyclopedia of Science Fiction*; Ash, Brian, Ed.; Harmony Books: New York, 1977.
3. Lem, Stanislaw. "The Time Travel Story and Related Matters of SF Structuring". In *Science Fiction: A Collection of Critical Essays*; Rose, Mark, Ed.; Prentice Hall: Englewood Cliffs, NJ, 1956.
4. Knight, Damon. *In Search of Wonder*; Advent: Chicago, IL; 1967.
5. Dick, Philip K. "Afterword to 'A Little Something for Us Tempunauts.'" In *Final Stage: The Ultimate Science Fiction Anthology*; Ferman, Edward L.; Malzberg, Barry N., Eds.; Charterhouse: New York, 1974.
6. Nicholls, Peter. "Metaphysics". In *The Science Fiction Encyclopedia*; Nicholls, Peter, Ed.; Doubleday: Garden City, NY, 1979.
7. Bretnor, Reginald. *Modern Science Fiction, Its Meaning and Future*; Coward-McCann: New York, 1953.
8. Pohl, Frederik. "Essay". In *The Visual Encyclopedia of Science Fiction*; Ash, Brian, Ed.; Harmony Books: New York, 1977.

9. Merril, Judith. "What Do You Mean: Science? Fiction?" In *SF: The Other Side of Realism*; Clareson, Thomas D., Ed.; Bowling Green University Popular Press: Bowling Green, KY, 1971. Rose, Mark, Ed.; Prentice Hall: Englewood Cliffs, NJ, 1976.

10. Conquest, Robert. "Science and Literature". In *Selected Fiction: A Collection of Critical Essays*; Rose, Mark, Ed.; Prentice Hall: Englewood Cliffs, NJ, 1976.

History and Tradition

★3★

The Search for a Definition of Mankind

H. G. Wells and His Predecessors

Ben B. Chastain

★In his interesting and provocative history of science fiction, *Trillion Year Spree* (1), Brian Aldiss proposes a working definition of the field that will serve well for this chapter: "Science fiction is the search for a definition of mankind and his status in the universe which will stand in our advanced but confused state of knowledge (science), and is characteristically cast in the Gothic or post-Gothic mode." It is distinguished from pure fantasy, then, by its use of (or at least mention of) current scientific knowledge as a basis for its imaginative developments. Using this definition, Aldiss makes a persuasive case (2) for beginning a study of modern science fiction with Mary Shelley's novel, *Frankenstein,* published in 1818.

In the Beginning

There were, of course, many worthy ancestors of science fiction; I will mention only two. The great German astronomer Johannes Kepler wrote a book entitled *Somnium* (Dream); it was published in 1634, four years after his death. In it his hero is transported to the moon by supernatural means; however, once he arrives there the description of the moon—its structure, climate, geography, and inhabitants—is strictly in keeping with the knowledge then recently obtained by telescopic observations. A much better known book is Jonathan Swift's *Gulliver's Travels*, published in 1726. Aldiss reluctantly refuses to admit this superb satire to his canon, but in the account of the visit to the flying island of Laputa we are given a detailed explanation of the use of magnetic attraction and repulsion to control the island's flight, complete with diagram. (3).

In the opening paragraph of the Preface to the first edition of *Frankenstein* (4), Mary Shelley gives us her proposed definition of the then nonexistent genre:

> The event on which this fiction is founded has been supposed, by Dr. [Erasmus] Darwin, and some of the physiological writers of Germany, as not of impossible occurrence. I shall not be supposed as according the remotest degree of serious faith to such an imagination; yet, in assuming it as the basis of a work of fancy, I have not considered myself as merely weaving a series of supernatural terrors. The event on which the interest of the story depends is exempt from the disadvantages of a mere tale of spectres or enchantment. It was recommended by the novelty of the situations which it developes; and, however impossible as a physical fact, affords a point of view to the imagination for the delineating of human passions more comprehensive and commanding than any which the ordinary relations of existing events can yield.

The scientific question was that of the nature of the life force, specifically whether galvanism (electricity) might be that force. His turning from the study of ancient science (Paracelsus and the alchemists) to that of modern science (chemistry and electricity) starts Victor Frankenstein on the road to his role as the "modern Prometheus" of the novel's subtitle. This comes about through the influence of a Professor Waldman, whose lecture is described in detail by Frankenstein (5):

> He began his lecture by a recapitulation of the history of chemistry, and the various improvements made by different men of learning. ... He then took a cursory view of the present state of the science, and explained many of its elementary terms. ... He concluded with a panegyric upon modern chemistry, the terms of which I shall never forget:

"The ancient teachers of this science", said he, "promised impossibilities and performed nothing. The modern masters promise very little; they know that metals cannot be transmuted, and that the elixir of life is a chimera. But these philosophers, whose hands seem only made to dabble in dirt, and their eyes to pore over the microscope or crucible, have indeed performed miracles. They penetrate into the recesses of nature, and show how she works in her hiding places. They ascend into the heavens; they have discovered how the blood circulates, and the nature of the air we breathe. They have acquired new and almost unlimited powers; they can command the thunders of heaven, mimic the earthquake, and even mock the invisible world with its own shadows."

Such were the professor's words—rather let me say such the words of fate, enounced to destroy me.

Waldman also advises him to study not just chemistry but "every branch of natural philosophy, including mathematics", and states that "A man would make but a very sorry chemist if he attended to that department of human knowledge alone."

M. K. Joseph summed up the importance of *Frankenstein* to science fiction in his Introduction (6):

Mary Shelley wrote in the infancy of modern science, when its enormous possibilities were just beginning to be foreseen by imaginative writers like Byron and Shelley and by speculative scientists like Davy and Erasmus Darwin. At the age of nineteen, she achieved the quietly astonishing feat of looking beyond them and creating a lasting symbol of the perils of scientific Prometheanism.

The book is much more subtle and profound than any of the movies made from it; if you have not read it, do so. Mary Shelley wrote at least one other novel that might be considered science fiction, *The Last Man*, published in 1826. It is set late in the 21st century and tells of a plague that destroys mankind.

After Shelley, the next well-known author who contributed to science fiction is Edgar Allan Poe. He is often referred to as the father of American science fiction; but, as Aldiss (7) put it, "Poe's best stories are not science fiction, nor his science fiction stories his best." Two examples will suffice. In "The Facts in the Case of M. Valdemar" (1845), the then-new science of mesmerism plays a central role: A sick man is hypnotized, and when he dies his soul cannot be released, nor can his body decay, until the hypnotic bond is broken. In "The Unparalleled Adventures of One Hans Pfaall" (1835), an alleged journey to the moon is made via balloon; the preparations and the trip are described in technical detail; the lunar world is glimpsed briefly. There is some chemistry involved:

> [I obtained] a quantity of a *particular metallic substance, or semi-metal,* which I shall not name, and a dozen demijohns of *a very common acid.* The gas to be formed from these latter materials is a gas never yet generated by any other person than myself … or at least never applied to any similar purpose. I can only venture to say here, that it is a *constituent of azote,* so long considered irreducible, and that its density is about 37.4 times *less than that of hydrogen.*

Science Fiction as a Genre

The first writer to systematically exploit the possibilities of science fiction, establish it as a genre, and make it a huge commercial success was Jules Verne (1828–1905) (8). In a long and varied series of novels known collectively as *Les Voyages Extraordinaires,* Verne celebrated 19th-century scientific progress. In books such as *Journey to the Center of the Earth* (1864), *From the Earth to the Moon* (1865), *Twenty Thousand Leagues Under the Sea* (1870), *Mysterious Island* (1875), and *Hector Servadac* (1877), he carried his readers on journeys both strange and exciting, while imparting enormous amounts of scientific information. For instance, in *Hector Servadac,* perhaps the least well-known of the novels mentioned, Verne's protagonists actually travel on a comet throughout the solar system. As the journey proceeds, the reader is instructed in physics and astronomy, including (but not limited to) the sizes, masses, periods, and then-known characteristics of all the planets and their satellites; the effects of diminished gravitational force and atmospheric pressure; the sizes and masses of French gold, silver, and copper coins; and a detailed geography of the area of the Mediterranean. Some of this information is well-integrated into the story (e.g., the coins being used for accurate weights and measures), but much of it comes in large undigested lumps, holding up (or at least not advancing) the narrative.

Verne's didacticism is both a strength and a weakness of his fiction. As I. O. Evans (9, 10) put it,

> Sheer romancing figures little in [Verne's] work, and satire and utopianization play only a minor part in it. His great aim was to impart infor-

mation through the medium of an exciting story, and to this end he described his heroes' achievements with almost a documentary precision.

So great was his interest in geography and general science that he may well have taken it for granted that his readers shared it. To some extent they may actually have done so.

When his publisher rejected an 1861 story about a castaway on a deserted island because of its lack of science and its "languid" characters, Verne responded by studying chemistry and visiting a chemical works, then years later revising the story into *The Mysterious Island.* In this classic trilogy, five Americans escape from a Civil War prison in a balloon and land on a desert island with one match, one grain of wheat, two watches, and the metal collar on their dog. From these few items and their scientific ingenuity they set up a pottery, an iron foundry and a munitions works; they produce a wheat field and a vegetable garden; build a fortified home and a sailing-boat; domesticate the island fauna; and, in short, produce what they hope might become the 38th state of the Union. As a climax, they meet up with Captain Nemo and his submarine from *Twenty Thousand Leagues Under the Sea.*

For some readers and critics, science fiction is important mainly because of its ability to make technological predictions. Using this criterion, Jules Verne must rank high on any list of writers. His tales included highly efficient aircraft and submarines, the exploration of subterranean caverns and lakes, the survival into modern times of prehistoric creatures (remember the coelacanth?), rocket-propelled guided missiles, artificial satellites, the construction of an artificial island, and weather control. He also contributed a sense of humor and idealism.

Verne's place as one of the founding fathers of science fiction is unquestioned. However, his fit to the Aldiss quote at the beginning of this chapter is an uncomfortable one. Verne was a very religious man; his vision of "mankind and his status in the universe" was essentially an optimistic one, accepting the status quo (including racial and national stereotypes). If humans tried to usurp the place of God (as, say, in *Frankenstein*), the result could only be catastrophe. His heroes were often rebels or outcasts from society, and he was anti-Imperial in an age of Imperialism, but he felt that the future of mankind was in the hands of the Almighty, and he was sustained by that faith. He did not agree with Darwin's views on natural selection; for him there was no apeman intermediate capable of evolving into a human. And in what may have been his last writing, a short story called

"The Eternal Adam" (published posthumously in 1910), the cataclysm that destroys civilization spares a few who, under Providence, survive to replenish the earth. Mankind abides. His "status in the universe" is secure.

One last predecessor should be mentioned, Robert Louis Stevenson (1850–1894). His short novel *The Strange Case of Dr. Jekyll and Mr. Hyde*, first published in 1886, has met the same fate as *Frankenstein*—its continuing fame rests mainly on movie adaptations, none of which do it justice. Dr. Henry Jekyll is convinced of the dual nature of man; through his scientific studies he is able to compound a drug that, in his words, "so potently controlled and shook the very fortress of identity" as to bring about the separation of those natures, with tragic results. Like Shelley's novel, it is a cautionary tale for those who would experiment with our "advanced but confused state of knowledge." Again, if you have not read it, do so.

The Life of H. G. Wells

At last we have reached our major subject, the archetype of modern science fiction: Herbert George Wells. If we accept Brian Aldiss's idea that "the search for a definition of mankind and his status in the universe" is the touchstone of the field, then no one has had greater influence on the development of science fiction than Wells. His "scientific romances" written nearly a century ago are still read today, with very little allowance having to be made for their 19th-century science. His success at technological forecasting is comparable to that of Verne, but his main concern was the future, and the eventual fate, of humankind.

Numerous biographical and critical studies of Wells are readily obtainable *(11)*, so I will consider only a few characteristic "romances", especially those containing some chemistry. However, a writer's own life experiences obviously influence his fiction, so a brief look at Wells's formative years is in order, and his own words may be used. In 1895, writing to an editor who had accepted a short story and had asked for some facts about the author, Wells said *(12)*,

> It's awfully good of you to go writing up a reputation for me, and I very gladly do what you ask of me. I was born at a place called Bromley in Kent, a suburb of the damnedest, in 1866, educated at a beastly little private school there until I was thirteen, apprenticed on trial to all sorts of trades, attracted the attention of a man called Byatt, Headmaster of Midhurst Grammar School, by the energy with which I mopped up

Latin—I went to him for Latin for a necessary examination while apprenticed (on approval, of course!) to a chemist there, became a kind of teaching scholar to him, got a scholarship at the Royal College of Science, S. Kensington (1884), worked there three years, started a students' journal, read abundantly in the Dyce and Foster Library, failed my last year's examination (geology), wandered in the wilderness of private school teaching, had a lung haemorrhage, got a London degree, B.Sc. (1889) with first- and second-class honours, private coaching, *Globe* turnovers, article in the *Fortnightly* (1890), edited an obscure educational paper, had haemorrhage for the second time (1893), chucked coaching, and went for journalism.

At the Royal College of Science, Wells spent his first year happily studying biology with Thomas H. Huxley, the defender and popularizer of Charles Darwin. The idea of evolution in general and of natural selection in particular is a seminal one in Wells's writing. His second year was spent on physics; an apparent incompatibility with the professor's personality and methods caused him to lose interest in the prescribed course and "read abundantly in the … Library". The third year brought the aforementioned failure. His love affair with science, begun in earnest at Midhurst, was not greatly affected by this setback; he continued to study on his own, reading in logic and psychology to obtain teaching diplomas, and in chemistry to qualify for his B.Sc. from the University of London. He kept abreast of the latest developments in science throughout his life, and wrote science articles for newspapers and magazines as well as his "scientific romances" (*13*).

Wells's Career as a Writer

His journalistic career had begun with the *Science Schools Journal* in 1886; he edited the first few issues, and over the next four years contributed a number of essays and stories to it, including, in May 1887, "A Tale of the Twentieth Century", in which perpetual motion is applied to the London Underground Railway with comically tragic results. Of greater importance, there appeared in the April, May, and June issues of 1888 the first three installments of his *The Chronic Argonauts*; this first novel was never completed (in his autobiography he said "That I realized I could not go on with it marks a stage in my education in the art of fiction."), but it turned out to be a sort of first draft for his first triumph, *The Time Machine* (*14*).

In *The Chronic Argonauts* his hero, one Moses Nebogipfel, Ph.D., F.R.S., etc., is rather like the stereotypical mad scientist; in his final incarnation as the Time Traveller he would become much more "normal". There is little chemistry as such in the story, though the preparations for the construction of the time machine include "crates filled with grotesquely contorted glassware", "vast iron and fire-clay implements of inconceivable purpose", and "jars and phials labeled in black and scarlet 'POISON'" (*15*). It also contains Wells's equivalent of Mary Shelley's disclaimer; in showing his machine to a clergyman who has confronted him, Dr. Nebogipfel says, "There is absolutely no deception, sir, I lay no claim to work in matters spiritual. It is a bona fide mechanical contrivance, a thing emphatically of this sordid world" (*16*).

In 1894, Wells published a short story called "The Diamond Maker" (*17*) in which the narrator meets a man who claims to have discovered a method for manufacturing artificial diamonds. In 1893, Henri Moissan, the French chemist and future Nobel laureate, best known for having isolated the element fluorine, had reported success in making artificial diamonds, using the high-temperature electric furnace he had developed. Moissan's work was later disproved, but Wells had obviously read of it. (The first proven synthesis of diamonds was by General Electric in 1955.) In 1895 came "The Remarkable Case of Davidson's Eyes" in which a scientist, stooping between the poles of an electromagnet in a thunderstorm, suddenly sees not the world around him but an island beach. The island proves to be a real place some 8000 miles away. The idea of a "kink in space" or parallel worlds was one to which Wells would return.

The year 1895 also saw the publication of the final version of *The Time Machine*, the work that made him famous. The novel makes little direct use of chemistry, but its concern for mankind's status in the universe requires at least brief comment. In the world of the year 802,701 A.D., all nonhuman animal life, all undesired plants, and all diseases have apparently been deliberately extinguished. Humanity has evolved into two new species—the Eloi, a childlike race living a seemingly Edenic existence amid the ruins of earlier civilization; and the Morlocks, apelike, living underground, tending great machines, and, it turns out, feeding on the Eloi. *The Time Traveller* theorizes that the distance between Capital and Labor, between haves and have-nots, may have resulted in this differentiation. Both biological and social Darwinism have acted on mankind, with tragic results (*18*).

Traveling even further into the future produces a view of the Earth in which the sun has grown larger and duller, mankind has disappeared (the predominant species seems to be a crustacean), and the end seems near. The ideas of thermodynamics, including entropy and the Second Law, were also at work in Wells's mind.

In 1896 came "The Plattner Story", in which we meet a secondary school teacher, a modern languages master, who also has to teach chemistry, geography, bookkeeping, and other subjects as needs arise. One of his students brings a greenish powder he claims to have found, and Mr. Plattner begins to analyze it. But, as the narrator says, "Plattner's practical chemistry was, I understand, temerarious." An explosion results, and the teacher vanishes, apparently "blown clean out of existence". A few days later he reappears with a story of having been in another world, one whose shadowy inhabitants are always watching our world. His story meets with general skepticism, but it is discovered that his whole anatomy has been reversed—his heart is now on his right side, he writes from right to left, and so on. Once again we have the concept of parallel worlds, or some "kink in space", a theme that Wells popularized and that has served science fiction writers well ever since (e.g., Isaac Asimov's *The Gods Themselves*) (19).

The Island of Dr. Moreau, also published in 1896, brings echoes of *Frankenstein* and perhaps of *Dr. Jekyll*. Moreau is a scientist, but one who considers himself above conventional ethics. He attempts to create a new man, not from bits of corpses but from animals, by means of vivisection. His fate is the same as his predecessors; he is destroyed by his Beast People. But he leaves the narrator, and the reader, with the realization that modern man still contains his ancestral bestiality, sometimes very near the surface. Prendrick, the narrator, who identified himself early in the novel as having studied biology with Thomas H. Huxley, ends by withdrawing from society, and spends his time (20) "surrounded by wise books. ... My days I devote to reading and to experiments in chemistry, and I spend many of the clear nights in the study of astronomy."

The 1897 novel was *The Invisible Man*. The opening is borrowed from the unfinished *Cosmic Argonauts*—a mysterious stranger arrives in a village and begins to conduct scientific experiments. Crates are unpacked, producing (21)

> bottles—little fat bottles containing powders, small and slender bottles containing coloured and white fluids, fluted blue bottles labelled *Poison*. ... The chemist's shop in Bramblehurst could not boast half so many ... the only things that came out of these crates besides the bottles were a number of test-tubes and a carefully packed balance.

This mysterious stranger, unlike Nebogipfel, is completely covered by clothing and bandages, and eventually proves to be quite invisible. He is a man named Griffin, who "won the medal for chemistry" at University College, as he reminds a former classmate. Kemp, the classmate, says, "But what deviltry must happen to make a man invisible?" The reply is, "It's no deviltry. It's a process, sane and intelligible enough" (22). He explains that his work in optical density, pigments, and refraction has enabled him to develop a process whereby a person can lower his refractive index to that of air, except for the red of blood and the coloring in hair (23). He has done this with a kind of radiation, an ethereal vibration, which he says is "not these Roentgen vibrations" (X-rays, discovered in 1895). He has also developed a drug that decolorizes the blood without affecting its functions (24). But his interest in his discovery seems to lie in the power it gives him over ordinary men, and, like Frankenstein and Moreau, he is destroyed by his creation.

In 1898 Wells published what may be his best-known work, *The War of the Worlds*. Even those who have never read anything by Wells are aware of his invaders-from-Mars story, aided by the Orson Welles radio broadcast and the George Pal movie. Its underlying theme is stated very near the beginning (25):

> And we men, the creatures who inhabit this earth, must be to [the Martians] at least as alien and lowly as are the monkeys and lemurs to us. The intellectual side of man already admits that life is an incessant struggle for existence, and it would seem that this too is the belief of the minds upon Mars. ... And before we judge of them too harshly we must remember what ruthless and utter destruction our own species has wrought, not only upon animals, such as the vanished bison and the dodo, but upon its own inferior races. The Tasmanians, in spite of their human likeness, were entirely swept out of existence in a war of extermination waged by European immigrants, in the space of fifty years. Are we such apostles of mercy as to complain if the Martians warred in the same spirit?

The invaders must use machines to move about and to fight because they are "bipeds with flimsy, silicious skeletons and feeble musculature". In fact, the narrator muses (26),

> To me it is quite credible that the Martians may be descended from beings not unlike ourselves, by a gradual development of brain and hands (the latter giving rise to the two bunches of delicate tentacles at last) at the

expense of the rest of the body. Without the body the brain would, of course, become a mere selfish intelligence, without any of the emotional substratum of the human being.

Here is, clearly, one possible path for the evolution of man.

The chemistry in *The War of the Worlds* is confined to the weapons with which the Martians attack mankind, a Heat-Ray and a poisonous Black Smoke. The generating mechanism of the Heat-Ray is never discovered; we are told that "The terrible disasters at the Ealing and South Kensington laboratories have disinclined analysts for further investigations." The Smoke combined with the moisture of the air and sank to earth as a powder. Analysis of the powder points to an unknown element, described at one point (27) as "giving a group of four lines in the blue of the spectrum", at another "with a brilliant group of three lines in the green", and "it is possible that it combines with argon to form a compound which acts at once with deadly effect upon some constituent in the blood." (The elements argon and helium had been discovered on earth only in 1895; they were identified by their spectral lines.) *The War of the Worlds* is considered by many critics to be Wells's finest "scientific romance"; it is still popular today both as a great adventure story and as a "search for a definition of mankind".

In 1899 Wells published the novel *When the Sleeper Wakes* and the novellette *A Story of the Days to Come*. These stories, set at the beginning of the 22nd century, describe a future society that can only be called a negative utopia. As Graham, the hero (and the Sleeper) of the novel, summarizes (28),

> So the magnificent dream of the nineteenth century, the noble project of universal individual liberty and universal happiness, touched by a disease of honour, crippled by a superstition of absolute property, crippled by the religious feuds that had robbed the common citizens of education, robbed men of standards of conduct, and brought the sanctions of morality to utter contempt, had worked itself out in the face of invention and ignoble enterprise, first to a warring plutocracy, and finally to the rule of a supreme plutocrat.

In these two tales Wells becomes explicit in his use of fiction as a tool for the reform of society; later novels such as Aldous Huxley's *Brave New World* and George Orwell's *1984* are in his debt.

Two short stories from 1901 are appropriate for our study. In "Filmer" we have an account of the invention of the first flying-machine (two years before the Wright brothers). Filmer, like Wells, attends the Science School at South Kensington and receives a B.Sc. from London University (with double first-class honors in mathematics and chemistry, rather better than his creator). He is seen lecturing to the Society of Arts on "rubber and rubber substitutes". His airship uses the expansion and contraction of balloons inside the frame to control lift, and he invents and manufactures "a new substance" for the elastic lining of the balloons, which also turns out to be useful for "the valves of a new oil-engine". He commits suicide before the final test, but his machine is a success.

"The New Accelerator" reports on the discovery by a Professor Gibberne of a drug that speeds up all processes of the central nervous system and so slows the perceived passage of time. As he describes it to a friend,

> "[I]n this precious phial is the power to think twice as fast, move twice as quickly, do twice as much work in a given time as you could otherwise do."
>
> "But is such a thing possible?"
>
> "I believe so. If it isn't, I've wasted my time for a year. These various preparations of the hypophosphites, for example, seem to show that something of the sort … even if it was only one and a half times as fast, it would do."

The drug turns out to be much more powerful than that, causing various problems that the story works out.

The novel published in 1901 was *The First Men in the Moon*. The journey is made possible by the invention of a substance called Cavorite (after its inventor); it is "a complicated alloy of metals and something new … called, I believe, *helium*, which was sent to him from London in stone jars." The substance is opaque to all forms of radiation, including gravity (29). The manufacture involves heating the substances in a furnace to fuse them and then allowing the mixture to cool very slowly (rather like the manufacture of diamonds in his earlier story.) After an early mishap, the material is obtained, and a glass sphere lined with it is prepared for the trip. The vessel contains (30)

> compressed foods, concentrated essences, steel cylinders containing reserve oxygen, an arrangement for removing carbonic acid and waste from the air and restoring oxygen by means of sodium peroxide, water condensers, and so forth.

The reason for the trip is not simply the adventure or the scientific interest; it is assumed there will be mineral wealth, "sulphur, ores, gold perhaps; possibly new elements."

The travelers arrive on the moon, and the descriptions of the lunar landscape and plant life are among Wells's most vivid writing. The novel is mainly a pure adventure story, but the final chapters give a description of Selenite society, and once again we see Wells grappling with possible futures for mankind. The moon creatures have evolved through biological conditioning into a society of highly differentiated functions (something akin to ants or bees), and their intellectuals have enormous heads and rudimentary bodies, rather like the Martians (31).

Because both Verne and Wells wrote of trips to the moon, their novels have often been used to contrast the two giants of early science fiction. In a much-quoted passage (32), Verne had this to say about Wells:

> His books were sent to me, and I have read them. It is very curious, and, I will add, very English. But I do not see the possibility of comparison between his work and mine … It occurs to me that his stories do not repose on very scientific bases. No, there is no rapport between his work and mine. I make use of physics. He invents. I go to the moon in a cannon-ball, discharged from a cannon. Here there is no invention. He goes to Mars [sic] in an airship, which he constructs of a metal which does away with the law of gravitation. That is all very well … but show me this metal. Let him produce it.

Some 30 years later, Wells had this comment on Verne (33):

> [T]here was a disposition on the part of literary journalists at one time to call me the English Jules Verne. As a matter of fact there is no literary resemblance whatever between the anticipatory inventions of the great Frenchman and [my] fantasies. His work dealt almost always with actual possibilities of invention and discovery, and he made some remarkable forecasts. The interest he invoked was a practical one; he wrote and believed and told that this thing or that could be done, which was not at that time done. He helped his reader to imagine it done and to realize what fun, excitement, or mischief might ensue. Many of his inventions have "come true". But these stories of mine … do not pretend to deal with possible things; they are exercises of the imagination in a quite different field.

In fact, many of Wells's inventions have "come true" as well. But his approach was much more like that of Mary Shelley; his science "however impossible as a physical fact, affords a point of view to the imagination for the delineating of human passions" (4).

Bergonzi (34) considers *The First Men in the Moon* to be Wells's "last genuine novel-length romance", and feels that his later works belong more to the "publicist and pamphleteer" than to the literary artist. It is true that his efforts to reform society do begin to intrude upon his imagination, but I believe that a number of his later works are worthy of at least brief mention here.

In a 1903 short story called "The Truth About Pyecraft", an obese man hears from a friend, "Our Western pharmacopaeia is anything but the last word in medical science. In the East, I've been told …", and prepares an Eastern prescription for Loss of Weight. The result is that he becomes so light he floats up to the ceiling. The friend (named Formalyn) chides him, "What you wanted was a cure for fatness! But you always called it _weight!_", and recommends lead inserts in his underwear.

In 1904 Wells published the novel *Food of the Gods*, wherein two scientists named Bensington and Redwood develop a theory about the nature of growth of all living things, and then a substance that greatly increases growth (they call it Herakleophorbia—"the nutrition of a possible Hercules"). The first part of the novel is Wells in top form, with deft characterizations and some nice satirical comments on scientists, but the later sections become more crudely didactic, with the giant children (representing the future of mankind) becoming mouthpieces for his sociological ideas.

In the Days of the Comet, published in 1906, tells of the great change wrought upon the nature of mankind by the collision of a gaseous comet with the earth. The main point of the novel is to describe a Wellsian utopia (including free love), but there is some chemistry used. While the comet is still far away, we are told that "the spectroscope was already sounding its chemical secrets, perplexed by the unprecedented band in the green" (35). As it nears, scientists predict it will produce "a great blaze of shooting stars. They might be of some unwonted color because of the unknown element that line in the green revealed" (36). The impact and its aftermath are described thus (37):

> They say it was the nitrogen of the air, the old *azote*, that in the twinkling of an eye was changed out of itself, and in an hour or so became a respirable gas, differing indeed from oxygen, but helping and sustaining

its action, a bath of strength and healing for nerve and brain. I do not know the precise changes that occurred, nor the names our chemists give them; ... only this I know—I and all men were renewed.

One of Wells's inventions that "came true" was described in his novel *The World Set Free*. As he reports in his autobiography (38),

[E]arly in 1914 I published a futuristic story ... in which I described the collapse of the social order through the use of "atomic bombs" in a war that began ... with a German invasion of France by way of Belgium. After this collapse there was to be a wave of sanity—a disposition to believe in these spontaneous waves of sanity may be one of my besetting weaknesses—and a wonderful council at Brissago (near Locarno!) was to set up the new world order.

His scientists learn how to speed up the natural rate of decay of certain isotopes and to begin a chain reaction that will continue to explode for many days once the bomb is dropped. Wells was aware of the work of Ernest Rutherford, Frederick Soddy, and others; the main purpose of the book, however, is to describe the new world order.

In 1923 Wells returned to the concept of parallel worlds in *Men Like Gods*. A few earthlings are transported through a "kink in space" into a very Earthlike world, but one in an advanced state of civilization, a Utopia. Once again the shape of a new world order is suggested, but also several contemporaries are openly caricatured—Arthur Balfour, the young Winston Churchill, and Sir Edward Marsh. The Utopians have chosen which animal and plant species to keep and which to eliminate; we are told that (39)

Many otherwise obnoxious plants were a convenient source of chemically complex substances that were still costly or tedious to make synthetically, and so had kept a restricted place in life. ... [Other plants] had been trained and bred to make new and unprecedented secretions, waxes, gums, essential oils, and the like, of the most desirable quality.

We also learn that "Gold evidently was cheap in Utopia. Perhaps they knew how to make it" (40). There is mention of a group of chemical students doing research on some question of atomic structure. But near the end, in discussing the 200 or 300 years of earthly scientific endeavor, a Utopian scientist says, "We have gone on for three thousand years now, and a hundred million good brains have been put like grapes into the wine-press of science. And we know today—how little we know" (41).

One last novel will be mentioned here: *Star Begotten*, published in 1937 when Wells was 70 years old. It tells of another invasion from Mars, this time not with heat rays and black smoke but with cosmic thought rays, which produce a mutation in certain children, resulting in a new breed of humans who will save the world. The story tells of one Joseph Davis and his wife Mary, who gives birth to one of these humans.

Until his death in 1946, H.G. Wells continued to write and work for the coming of a "wave of sanity" that could save the world. If much of his later writing is polemical and prophetical, long out of print and hard to obtain, his "scientific romances", especially the early ones, are still very much alive. They continue to bring pleasure and to stimulate thought on "the search for a definition of mankind and his status in the universe".

References and Notes

1. Aldiss, Brian. *Trillion Year Spree*; Atheneum Publishers: New York, 1986. Highly recommended as a guide to the field.
2. Ibid., Chapter 1.
3. Swift, Jonathan. *Gulliver's Travels*; Part III, Chapter iii.
4. Shelley, Mary. *Frankenstein*. Not all modern editions include the 1818 Preface. This is quoted from an edition with an Introduction by M. K. Joseph, Oxford University Press: New York, 1969.
5. Ibid., Chapter III.
6. Ibid., p xiv.
7. Reference 1, Chapter II.
8. An excellent brief guide is by I. O. Evans, *Jules Verne and His Work*, Twayne Publishers: New York, 1966.
9. Ibid., p 155.
10. Ibid., p 150.
11. Wells's own *Experiment in Autobiography* (Macmillan: New York, 1934) is highly readable and informative. The standard biography is the one by Norman and Jeanne MacKenzie (Simon and Schuster: New York, 1973). J. R. Hammond, *An H. G. Wells Companion*, Barnes & Noble: New York, 1979, is a very useful guide to his fiction. A fine critical study of the scientific romances is by Bernard Bergonzi, *The Early H. G. Wells*, University of Toronto Press: Toronto, Canada, 1961.
12. Letter to Grant Richards, quoted in Bergonzi (ref. 11), p 23.
13. Many of Wells's science articles are collected in *H. G. Wells: Early Writings in Science and Science Fiction*; Philmus, R. M.; Hughes, D. Y., Eds.; University of California Press: Berkeley, CA, 1975.
14. "A Tale of the Twentieth Century" and *The Chronic Argonauts* are reprinted as an Appendix to Bergonzi's book (ref. 11).
15. Reference 11, p 192.
16. Ibid., p 207.

17. Wells's novels and stories appear in so many editions, both hardbound and paperback, that references to the novels will be given only by chapter. All of the short stories discussed here, except the first and the last, may be found in the paperback *Best Science Fiction Stories of H. G. Wells,* Dover Publications: New York, 1966. Dover also has published *Seven Science Fiction Novels of H. G. Wells* (hardbound) and *3 Prophetic Science Fiction Novels of H. G. Wells* (paperback).
18. *The Time Traveller,* Chapter VIII.
19. Reference my Asimov paper in this volume, "Beryllium, Thiotimoline, and Pâté de Foie Gras: Chemistry in the Science Fiction of Isaac Asimov".
20. *The Island of Dr. Moreau,* Chapter XXII.
21. *The Invisible Man,* Chapter III.
22. Ibid., Chapter XVII.
23. Ibid., Chapter XIX.
24. Ibid., Chapter XX.
25. *The War of the Worlds,* Book I, Chapter 1.
26. Ibid., Book II, Chapter 2.
27. Ibid., Book II, Chapter 10.
28. *When the Sleeper Wakes,* Chapter XIV.
29. *The First Men in the Moon,* Chapter I.
30. Ibid., Chapter III.
31. Ibid., Chapter XXIV.
32. *T.P.'s Weekly,* October 9, 1903.
33. *Collected Edition of the Scientific Romances,* London, 1933, Preface.
34. Reference 11, p 20.
35. *In the Days of the Comet,* Book I, Chapter 1.
36. Ibid., Book I, Chapter 5.
37. Ibid., Book II, Chapter 2.
38. Reference 11, p 569.
39. *Men Like Gods,* Book I, Chapter 6.
40. Ibid., Book I, Chapter 8.
41. Ibid., Book III, Chapter 3.

★4★

Planetary Chemistry in 100 Years of Science Fiction

Mark A. Nanny

★On December 7, 1995, we had our first direct contact with an outer planet as the *Galileo* probe descended into the Jovian atmosphere. Results showed that Jupiter's upper atmosphere contains a variety of elements and compounds: neon, argon, hydrogen sulfide, ammonia, methane, and hydrocarbons (1). In addition, a high abundance of krypton and xenon was discovered, and the water-vapor concentration was much lower than expected. The *Galileo* probe is our first contact with the surface of one of the outer planets, an event long dreamed and written about.

Since the 1890s, tales of planetary travel and exploration have been retold countless times with a myriad of plots. The planets of the solar system have been discovered and rediscovered within plots ranging from pure fantasy to meticulous detail based on contemporary scientific information. Some of these stories contain details about the planet's atmosphere, environment, and geochemistry in order to create a richer and more alien setting. Often, the information is scant, with only a brief mention of the surroundings, and thus, only casual inferences about the planet's chemistry can be made. However, in a few instances the descriptive chemical data are detailed enough to provide interesting and often insightful statements about planetary chemistry.

Although I will not attempt to be comprehensive, my intent in this chapter is to explore the atmospheres, environments, and geochemistry of planets in our solar system as presented in science fiction from the 1890s through the mid-1990s. Therefore, I have excluded works dealing with other worlds, natural disasters, and fantasy in order to focus on planetary science fiction containing chemical information.

Science fiction dealing with the planets and their exploration can be divided into four general periods that closely parallel advances in planetary science and exploration.

- The beginnings of planetary science fiction ranged from the 1890s through the 1910s. Most science fiction during this period relied heavily on fantasy and adventure rather than science.

- The second period spanned from the 1920s through the 1950s. The genre of this period is often referred to as space opera; the technology of space travel and the planetary environments are more detailed and advanced, but the plots usually are still based on adventure and fantasy.

- The work of the 1960s through the early 1970s consists of science fiction influenced by postwar technological developments, the beginning of the "space race", manned landings on the Moon, and unmanned probes flying by Mercury, Venus, and Mars. During this time, science fiction shifted to more realistic situations and viable technology.

- The final and current period (1970s through the 1990s) is marked by probe landings on Venus and Mars, in addition to detailed observation of the outer planets by unmanned spacecraft. The science fiction of this period begins to seriously incorporate accurate planetary chemical information, providing realistic and plausible stories of planetary exploration, colonization, terraforming, and the search for extraterrestrial life.

The Beginning of Planetary Science Fiction: 1890s–1910s

Planetary travel and exploration were rampant in science fiction from the 1890s through the 1910s; unfortunately, lack of detailed planetary scientific information caused most writings to be nothing more than adventure or

Significant Events in Planetary Science, 1600–1920

1610	Galileo (1564–1642) discovered four of Jupiter's moons: Io, Europa, Ganymede, and Callisto.
1610	Galileo first observed the rings of Saturn, but because of insufficient telescope power, he was unable to discern their true structure.
1645	Map of Venus created by F. Fontana.
1655	Christiaan Huygens (1629–1695) discovered Titan.
1659	Christiaan Huygens wrote *Systema Saturnium*, in which he correctly deduced the nature of Saturn's rings.
1664	Robert Hooke (1653–1703) observed the great red spot of Jupiter.
1666	Giovanni Domenico Cassini (1625–1712) observed white caps at the poles of Mars.
1682	Edmond Halley (1656–1742) observed "Halley's Comet" and calculated its 76-year revolution period.
1781	William Herschel (1738–1822) discovered Uranus.
1787	William Herschel discovered Titania and Oberon, two moons of Uranus.
1802	William Herschel glimpsed Umbriel, another moon of Uranus.
1845	John Couch Adam (1819–1892) mathematically predicted Neptune's position.
1846	Johann Galle and Heinrich D'Arrest (1822–1875) identified Neptune.
1846	William Lassell (1799–1880) discovered Triton, a moon of Neptune.
1877	Giovanni Virgino Schiaparelli (1835–1910), through careful study of Mars, observed numerous lines crossing large ochre areas of the planet's surface. He named these lines *canali*, which is Italian for canals.
1877	Asaph Hall (1829–1907), an American astronomer, discovered the moons Phobos and Deimos around Mars.
1881–1889	Giovanni Virgino Schiaparelli (1835–1910) mapped the surface of Mercury. It was determined later that streaks he observed on Mercury's surface were an optical illusion.
1896	Maurice Loewy (1833–1900) and Pierre Henri Puiseux (1855–1928) produced the first good photographic atlas of the Moon.

Source: See references 18, 23, 29, and 47.

fantasy stories set in exotic locations. Because little was known about the planet's surface, artistic license was used liberally to create environments resembling one of Earth's biomes, such as in John Jacob Astor's *A Journey in Other Worlds* (2), where the surface of Jupiter resembles a prehistoric jungle on Earth, or Gustavus W. Pope's *Wonderful Adventures on Venus* (3), where Venus is also similar to prehistoric Earth. Other stories describe utopian or dystopian, advanced or primitive civilizations in an Earthlike environment on Mars [*Across the Zodiac: The Story of a Wrecked Record* (4) by Percey Greg, *A Plunge into Space* (5) by Robert Cromie, *A Journey to Mars* (6) by Gustavus W. Pope, *Two Planets* (7) by Kurd Lasswitz, *The Certainty of a Future Life on Mars* (8) by Louis Pope Gratacap, *To Mars via the Moon* (9) by Mark Wicks, and *A Princess of Mars* (10) by Edgar Rice Burroughs], or Venus [*To Venus in Five Seconds* (11) by Jane Fredrick Thomas and *A Columbus of Space* (12) by Garrett P. Serviss].

These stories were primarily romantic adventures or social fantasies, but other science fiction was being written at this time that had some technical content in addition to the large amount of romantic adventure. Such works were often called "scientific romances". Jules Verne's *From the Earth to the Moon* (13) and Garrett P. Serviss's *The Moon Metal* (14) are examples. H. G. Wells's *The First Men in the Moon* (15) presents fantastic things such as "negative gravity" while at the same time hinting of future technology such as "wireless messages".

Despite the lack of accurate scientific content in most science fiction of this period, *The Doom of London* (16) by Robert Barr is unique. Even though the story occurs on Earth, it stands out in its description and effects of the atmospheric phenomenon of smog, formed from coal-combustion smoke and fog.

> London at the end of the nineteenth century consumed vast quantities of a soft bituminous coal for the purpose of heating rooms and of preparing food. In the morning and during the day, clouds of black smoke were poured forth from thousands of chimneys. When a mass of white vapor arose in the night, these clouds of smoke fell upon the fog, pressing it down, filtering slowly through it, and adding to its density. ... Once this condition prevailed, nothing could clear London but a breeze of wind from any direction.

In *The Doom of London*, the smog conditions last for a week, after which everyone in London dies from the "deadly effect of the deoxygenized atmosphere". The real health hazards of smog, not mentioned by the author, are the formation of suspended droplets of sulfuric acid, the presence of sulfur dioxide, and a variety of suspended particles. The main character is saved by a device that emitted oxygen gas, given to him by an eccentric American inventor. Using this device, he eludes death and escapes to the English countryside. The details in *The Doom of London* quite vividly foretold of smog disasters that occurred decades later in London and in Pittsburgh, Pennsylvania.

Space Opera: 1920s–1950s

During this period, space travel and planetary descriptions in science fiction became more realistic as technology advanced from simple airplanes to the development of rockets. Science fiction was rapidly evolving in its depth and sophistication; instead of prehistoric jungles and exotic societies, planets were now visited by rocket ships, and space suits were necessary apparel. Even so, much liberty was taken in describing planetary surfaces and the corresponding atmosphere, environment, and geochemistry, not to mention the unusual and unrealistic flora and fauna.

Mars was one of the most heavily visited planets during the space opera period, and hence, a myriad of Martian landscapes have been described. In 1934, Stanley G. Weinbaum, in *A Martian Odyssey* (*17*), has the main character Dick Jarvis, who is an astronaut chemist, hike for several days across Mars. Weinbaum presents the surface and atmosphere of Mars as harsh, dry, and desertlike, but viable for humans. Dick Jarvis can breathe the thin atmosphere and has to worry only about freezing temperatures after sunset. He has a special insulated sleeping blanket to keep him warm at night, but yet he still manages to get his nose frost-bitten. In reality, the Martian atmosphere consists of 95.32% CO_2 and only 0.13% O_2 by volume; in contrast, Earth's atmosphere is 78.1% N_2, 20.9% O_2, 0.9% Ar, and 0.03% CO_2 (*18*). In addition, the atmospheric pressure of Mars is less than 1/100 of Earth; hardly a hospitable atmosphere for humans (*18*).

Along his journey he encounters various life-forms, one of which is silicon-based. This creature consists of a "body like a big gray cask, an arm, and a sort of mouth-hole at one end; [and a] stiff pointed tail at the other." Jarvis tells his colleagues that this creature burrowed its "tail into the sand, pushed itself upright, and just sat" consuming "pure silica in the sand" and

Significant Events in Planetary Science, 1920–1950

1920	Walter Baade (1893–1960) discovered the asteroid Hidalgo.
1924–1929	Eugenios M. Antoniadi (1870–1944) mapped Mercury.
1929	Bernard Lyot (1897–1952) examined polarization of light from Mars and showed that Martian atmosphere is affected by haze or dust.
1930	Clyde Tombaugh (1906–) discovered Pluto.
1940	Walter Baade discovered the asteroid Icarus.
1944	Gerard P. Kuiper (1906–1973) discovered that Titan had an atmosphere.
1948	Gerard P. Kuiper at McDonald Observatory discovered Miranda, one of Uranus' moons.

Source: See references 18, 23, 29, and 47.

then, ten minutes later, defecated a silicon oxide "brick". How this creature found any chemical energy generating a silicon oxide brick from silica sand (i.e., silicon oxide) is a thermodynamic mystery.

Walter M. Miller, Jr., presents a slightly more realistic Martian environment in *Crucifixus Etiam* (*19*). The story follows Manue Nanti, who becomes a manual laborer on Mars for money and travel. Miller ignores problems with freezing temperatures but does recognize that Mars has a low oxygen concentration. Even so, the atmosphere he presents is more hospitable than the true Martian atmosphere. In *Crucifixus Etiam*, the pressure of the Martian air is so low that breathing is achieved using a "mechanical generator [that] served as a lung, sucking blood through an artificially grafted network of veins and plastic tubing, frothing it with air from a chemical generator, and returning it to his circulatory system." In addition to the atmosphere, Miller mentions Martian geochemical topics of interest such as lakes of "nearly pure iron-rust [that] was scooped into a smelter and processed into various grades of steel" and great reservoirs of tritium ice.

Arthur C. Clarke's *The Sands of Mars* (*20*) presents Mars with a thin atmosphere rich in CO_2, but one in which a person can survive with only a simple oxygen mask. Clarke presents Mars several years after colonization, and several small, dome-covered cities and research stations now exist. The ultimate goal of the inhabitants is to terraform Mars into an environment hospitable for humans. Oxygen for the cities' atmosphere is obtained by extracting oxygen from metallic oxides, primarily iron oxide, found in the Martian soil, and the CO_2 in the dome-covered cities is removed by plants brought from Earth. The excess CO_2 of the Martian atmosphere induces rapid plant growth, producing giant, towering flowers and shrubs. Outside the covered cities, Martian plants abound in diversity and form. Despite the high atmospheric CO_2 levels, Clarke surprisingly has these Martian plants obtain all of their carbon, in addition to minerals, from the soil and their energy from sunlight. These plants "can grow in a complete vacuum, like the plants on the Moon, if they've got suitable soil and enough sunlight."

One unique Martian plant has the ability to extract and concentrate oxygen from metallic oxides present in the soil and store it in "numerous 'pods' in the seaweed-like fronds … under quite high pressure". When these pods are broken open, the oxygen bubbles out of a brown, gooey liquid. Large Martian herbivores, which look like "plump kangaroos", obtain all their oxygen and food requirements by grazing on these plants. Despite the fantastical flora and fauna presented in *The Sands of Mars*, this story is unique in that it begins to seriously examine problems involved in terraforming an alien planet.

In 1957, Robert Silverberg described the surface of Mercury in *Sunrise on Mercury* (21). Silverberg presents Mercury with a orbital rotation that is synchronous with its axial rotation, therefore, one-half of the planet constantly faces the sun while the other half is in perpetual darkness. It was not until 1962 that W. E. Howard and his colleagues, using radar, measured that the axial rotation of mercury was two-thirds that of its orbital period. Thus the entire surface of Mercury eventually is heated by the sun (18).

Because Mercury's rotation was not understood when Silverberg wrote *Sunrise on Mercury* [a point that he makes in the introduction of his anthology *Tomorrow's Worlds* (22)], he divided Mercury into two distinct fractions: "the brightness of Sunside, the unapproachable inferno where zinc ran in rivers, and the icy blackness of Darkside, dull with its unlit plains of frozen CO_2." In reality, the sunside and darkside temperatures of Mercury are >600 K and <90 K, respectively (23). Any frozen CO_2 would quickly evaporate once on Sunside, especially with a surface pressure of 10^{-12} millibar, and once evaporated, very little would remain associated with the planet because Mercury does not have enough mass to hold an atmosphere. Extremely low levels of helium are present, though, presumably from the beta-decay of uranium and thorium or captured from the solar wind by Mercury's magnetic field (23).

The fact that Silverberg mentions "rivers of molten zinc" also indicates that this story was written before Mercury was well understood. According to the equilibrium condensation model for planet formation (24, 25) and other observational scientific data, Mercury is believed to consist of 70% metallic iron, comprising the core, and 30% silicates and magnesium, comprising the crust (18).

Jupiter and Uranus were two of the outer planets visited in science fiction writings during this period. The story *Desertion* (26) by Clifford D. Simak, presents the harsh Jovian surface and atmosphere with alkaline rains complete with liquid ammonia and hydrogen. This description is not too far-fetched because the atmosphere of Jupiter is believed to be, by volume, ~90% hydrogen, ~4.5% helium, and ~0.02% ammonia (2, 6). On the other hand, the fact that he describes a viable and distinct planetary surface is probably erroneous because Jupiter has a low density and is believed to be composed primarily of hydrogen and helium in the form of a gas or liquid (18, 27). In fact, many are doubtful that Jupiter even has a distinct surface. Simak also describes the Jovian surface as containing majestic cascades of liquid ammonia flowing over solidified oxygen! This is most surprising, for oxygen has a much lower melting point than ammonia; either the ammonia would freeze upon contact with the solid oxygen, or the oxygen would quickly melt and boil away. At Earth's atmospheric pressure, the melting point of O_2 is –218.4 °C and NH_3 is –74 °C. Also surprising is the fact that such large amounts of pure oxygen are present in the reduced atmosphere of hydrogen and ammonia, especially in the presence of the numerous lightning bolts mentioned. In this situation the oxygen and hydrogen would quickly react to form water.

Uranus is another planet where great artistic liberty was taken by Stanley G. Weinbaum in *The Planet of Doubt* (28). Weinbaum has several explorers land on a solid surface with a foggy, argon-rich, and oxygen-containing atmosphere. This atmosphere is viable for humans, and the astronauts can explore without space suits or breathing apparatus. Not only is Uranus warmed to comfortable temperatures by heat from its interior, but it is rich in unusual life-forms. To be fair to Weinbaum, Uranus is a mysterious planet, presenting a featureless surface, and information has only recently been obtained about its composition and structure. Uranus is believed to have more methane than Jupiter or Saturn and is thought to consist of three layers: a rocky core mainly composed of iron and silicates; a mantle of water, ammonia, and methane; and finally, a low-density surface of hydrogen, helium, and methane, which gives rise to its bluish-green color (29). Its only energy source is the sun, creating an effective surface temperature of 57 K, hardly a place for life-forms and explorations in casual clothing.

Postwar Technology and The Space Race: 1950s–Early 1970s

On October 4, 1957, the Soviet Union started the space race by successfully launching *Sputnik* into Earth's orbit. Afterwards, the Moon became the desired goal. The Soviet Union obtained photographs of the far side of the Moon with *Luna 3* in 1959, and the United States landed on the Moon in 1964 and 1965 with the unmanned *Ranger* probes. In 1968, American astronauts successfully circled the moon in *Apollo 8*, and on July 20, 1969, Neil Armstrong (*Apollo 11*) became the first person to set foot on the Moon. These lunar voyages, in addition to numerous unmanned probes orbiting Mars (*Mariner* probes, 1964–1971) and Venus (*Mariner* probes, 1962–1973, and *Venera 4*, 1967, the Soviet probe that actually descended for 94 minutes into the Venusian atmosphere) provided a wealth of scientific data about interplanetary travel and the composition and structure of the surface and atmosphere of the Moon, Mars, and Venus. This data influenced the detail and settings of planetary science fiction. Stories became more realistic in their portrayal of the surface conditions, geology, chemistry, and meteorology. Gone were the romantic days of astronauts jumping unprotected out of their spacecraft and finding bug-eyed aliens. Explorers now faced poisonous atmospheres of hydrogen, methane, and ammonia; extremely cold temperatures; and serious quests for the presence of life on other planets.

Arthur C. Clarke opened this period with an exploration of Venus in *Before Eden* (30). Scientists exploring near the south pole of Venus discover that temperatures are low enough to permit liquid water to exist. In addition, they encounter plant life that resembles a "velvet carpet", flowing like a "tide, or creeping carpet". In the muck of the "green twilight", it appears black, but in the beam of the scientist's flashlight, it is composed of "glorious vivid reds, laced with streaks of gold." This plant life is similar to that on Earth in that it uses atmospheric CO_2, converting it into the waste product O_2. This fact is pointed out in the story by noting that the atmosphere of Venus is mostly carbon dioxide, ammonia, and methane, except at the south pole, where the oxygen concentration reaches 15 parts per billion.

Significant Events in Planetary Science, 1950–1970s

1957, Oct. 4	*Sputnik* (U.S.S.R.) launched successfully into Earth's orbit.
1962	W. E. Howard and colleagues, using radar measurements, disproved that Mercury has a synchronous axial rotation.
1965	*Mariner 4* (U.S.) sent back close-up pictures of Mars' surface.
1971	*Mariner 9* (U.S.) began to systematically map Mars.
1972–1973	*Pioneer 10* and *Pioneer 11* (U.S.) did a flyby of Jupiter and Saturn.
1973–1974	*Pioneer 10* and *Pioneer 11* mapped out Jupiter's intense magnetic and gravity fields.
1974	*Mariner 10* (U.S.) did a flyby of Mercury and obtained surface information.

Source: See references 18, 23, 29, and 47.

Clarke's portrayal of the Venusian surface is quite advanced considering that until 1962 very little was known about Venus because of its thick cloud cover that prevented any photographic studies. Before 1962, it was believed that the clouds were water vapor and that the Venusian surface was covered with oceans. The findings of *Mariner 2* (1962) destroyed this oceanic theory of F. L. Whipple and D. H. Menzel (*18*). Other probes eventually mapped 80% of the surface, and Venus is now known to have a static crust, consisting mostly of huge rolling plains, some depression areas, and two high-land areas (*18*).

Frederik Pohl in *The Merchants of Venus* (*31*) created a hot Venusian surface with an atmosphere rich in CO_2 with a 20,000-millibar pressure at the surface. He also mentions upper atmospheric clouds containing hydrochloric acid and hydrofluoric acid and colors the lower clouds in the Venusian atmosphere with the presence of red mercuric sulfide and mercurous chloride.

One thermodynamically interesting point Pohl makes is the reduced efficiency of engines operating in an extremely hot environment. Pohl's main character, a captain of a surface shuttle, explains to a young lady whom he is trying to impress that the Carnot efficiency of an engine "is expressed by its maximum temperature—the heat of combustion, let's say—over the temperature of the exhaust. Well, the temperature of the exhaust can't be less than the temperature of what it flows into—otherwise you're not running an engine, you're running a refrigerator."

What is known about Venus has been obtained mainly by unmanned probes. These probes have shown that the atmosphere of Venus is composed, by volume, of 97% carbon dioxide, 0.1% water vapor, 0.005% carbon monoxide, 6×10^{-5} % hydrochloric acid, and 5×10^{-7} % hydrofluoric acid, in addition to nitrogen and a host of sulfur compounds: sulfur dioxide, carbonyl sulfide, hydrogen sulfide, and sulfuric acid (*2, 7*). The atmospheric composition is responsible for a portion of the high surface temperature of Venus, which is a uniform 737 K. These gases cause a greenhouse effect to occur; carbon dioxide accounting for 55% of the trapped heat, and water vapor and sulfur dioxide accounting for 25% and 5%, respectively. The remaining 15% is believed to be due to clouds and haze (*18*).

In *A Meeting with Medusa* (*32*), Arthur C. Clarke has Howard Falcon explore the Jovian atmosphere for two days in a heated hydrogen balloon. The upper atmosphere is portrayed accurately as containing hydrogen and helium, with cirrus clouds composed of ammonia crystals. As Falcon descends, he uses infrared (IR) spectroscopy to detect "complex carbon compounds", hydrogen, helium, and ammonia. He comments that "half of the basic molecules for life" are present! Further down, he encounters a "witch's cauldron of hydrocarbons" and "petrochemicals", not to mention being in a hydrocarbon "snowstorm".

Titan, one of Saturn's moons, is presented in James Blish's *How Beautiful with Banners* (*33*) as a frozen, dark place containing a thin methane atmosphere and covered with methane snow. Dr. Ulla Hillstrom, encased in a "virus bubble" space suit, a suit that is a single protein molecule and has the capability to maintain constant conditions, encounters strange life-forms that eventually cause her to meet her demise. The use of Titan is interesting because even though methane is not the major atmospheric constituent, it is the major organic constituent in the atmosphere and is present by volume at 0.01%. Nitrogen is ~94% by volume of Titan's atmosphere, and helium makes up the remaining 6%. Many simple organic compounds are present in Titan's atmosphere: ethane, acetylene, propane, diacetylene, hydrogen cyanide, cyanoacetylene, and cyanogen. It is believed that free methane is irreversibly converted into ethane and acetylene, both of which are present at 4- and 150-times greater concentration, respectively, than on Saturn. Titan is also chemically unique in that it contains ethylene, which has not been detected on Saturn, and that it also has hydrogen cyanide, which has not been detected on any of the other planets except Earth. Hydrogen cyanide is important because it is a key intermediate between the synthesis of amino acids and bases present in nucleic acids, giving rise to speculations of life on Titan (*18, 29, 34*).

The two farthest planets from the sun, Neptune and Pluto, are included in *One Sunday in Neptune* (35) by Alexei Panshin and *Wait It Out* (36) by Larry Niven, respectively. In *One Sunday in Neptune*, a group of bored men stationed at a base on Neptune's moon, Triton, decide to build a bathyscaph supported by a hydrogen-filled balloon and then sail to Neptune to explore its atmosphere. The story mostly accounts the building of the bathyscaph and their journey through Neptune's clouds of ammonia and methane, including an ammonia snowstorm. This story is unique in that it recognizes the current thinking regarding Neptune's structure. Because of its low density, Neptune is thought to consist of only gases, only liquids, or both (29). One character exclaims that "[a]t one time, it was expected to have a layer of ice and a rocky core beneath its atmosphere. In fact however, it has no solid surface. It's all atmosphere, a murky green sea of hydrogen and helium and methane and ammonia." Beyond this, very little is known about Neptune. It is known to have a reducing atmosphere of hydrogen, helium, methane, and possibly ammonia. Its low density suggests a composition of elements such as oxygen, nitrogen, silicon, and iron. Unlike Uranus however, it does emit heat from an unknown internal process (29).

Larry Niven's *Wait It Out* is an unusual story about an astronaut who becomes stranded on Pluto and decides to end it all by allowing himself to freeze into a heroic pose. The problem is that although his body is frozen solid, his mind still works, albeit very slowly, and he is cognizant of everything around him. Niven correctly portrays Pluto as a rocky, ice-laden planet covered with the "dim gray-white of snow". Due to Pluto's low density, it is believed that ice is a major component. The solid surface is described by Niven as frozen gases covering "water-ice mixed haphazardly with lighter gases and ordinary rock". Pluto's polar caps consist of methane ice; its thin atmosphere, with a pressure 1/100,000 of Earth's, contains methane and probably carbon monoxide and nitrogen; and its surface temperature is ~43 K (18, 29).

Planetary Probe Landings and Contemporary Times: 1970s–1990s

The amount of planetary information obtained in the past 16 years has been greater than any other time in history, and this clearly has affected science fiction. Numerous probes have landed on Venus [the Soviet

Significant Events in Planetary Science, 1970s–1990s

1970–1981	*Venera 7* through *Venera 14* (U.S.S.R.) visited Venus.
1975	*Viking 1* and *Viking 2* (U.S.) visited the planetary surface of Mars.
1976, July 20	*Viking 1* landed at Chryse Planitia, Mars.
1977	*Voyager* (U.S.) probes 1 and 2 launched.
1977	Uranus discovered to have a complex system of at least nine rings.
1977–1989	*Voyager* probes do a flyby of Jupiter and her moons (1979), Saturn and her moons (1980–1981), Uranus (1986), and Neptune (1989).
1978	*Pioneer Venus* (U.S.) visited Venus.
1979, March	*Voyager 1* observed erupting volcanoes on Io.
1979, July	*Voyager 2* observed a second erupting volcano and a caldera (volcanic opening) on Io.
1979	*Voyager 1* discovered that Jupiter has a faint ring.
1995	Northern polar cap and surface features detected on Pluto by Hubble telescope.
1995	Extrasolar system planets discovered orbiting the stars 51 Pegasi and GL229.
1995	*Galileo* spacecraft observed Jupiter and probe entered upper Jupiter atmosphere.

Source: See references 18, 23, 29, and 47.

Union's *Venera 7–14* (1970–1981) and the United States' *Pioneer Venus* (1978)] and on Mars [*Viking 1* and *Viking 2* (1975)]. Likewise, vast amounts of IR, visible, and ultraviolet (UV) spectroscopic data have been collected about the outer planets with the fly-by of Jupiter and Saturn with the probes *Pioneer 10* (1972) and *Pioneer 11* (1973) and the *Voyager 1* and *Voyager 2* probes. Both *Voyager* probes were launched in 1977 and passed Jupiter and its moons, Saturn and its moons, Uranus, and Neptune over the period of 1977 to 1989. More recently, the Hubble space telescope has been put into Earth's orbit, greatly enhancing observational capabilities, and the *Galileo* probe has observed Jupiter and entered its upper atmosphere.

Using data gathered from many of these missions, Arthur C. Clarke takes us on a spectacular voyage to Jupiter and her moons in *2010: Odyssey Two* (37) and *2061: Odyssey Three* (38). Both novels are extensions of his science fiction classic *2001: A Space Odyssey* (39). *2010: Odyssey Two* opens with a joint American and Soviet effort to reach the ship *Discovery*, which was left in orbit around Jupiter at the end of *2001: A Space Odyssey*. Once en route to Jupiter, they become aware that they are in competition with a secret Chinese mission also trying to reach the *Discovery*. The Chinese have designed their ship to carry only enough propellant for a one-way trip; they stop first on Europa to refill their propellant tanks with water for their muon-catalyzed fusion drives. Clarke has all the ships in these novels propelled by these devices; he explains that muon-catalyzed fusion was first discovered in the 1950s, but it was not until 2040 that stable muonium–hydrogen compounds were unexpectedly and accidentally synthesized. This discovery made small, portable nuclear power plants feasible. A working fluid was still required to provide thrust, and water turned out to be the cheapest, cleanest, and most convenient fluid. Thus, any planet or moon containing water could be used as a fueling site. Clarke never mentions how ships in 2010 could use a propulsion device developed in 2040.

From space, Clarke describes Europa as appearing uniformly pink with a few small brown patches and an intricate network of narrow lines, curling and weaving in all directions. Europa is further described as having a smooth ice surface, up to several kilometers thick, under which liquid water exists for up to 50 kilometers. The brown patches and narrow lines are areas where the ice cracked apart and refroze. The water underneath the surface is warmed by heat from tidal forces acting on Europa from Jupiter and the other moons. The tidal forces provide enough energy so that the ice surface is continually melting, breaking up, and freezing. Beyond this point, Clarke engages in fascinating speculation regarding Europa's biota and its chemical basis.

While the Chinese ship *Tsien* is refueling on the Europian surface, it is destroyed by a giant Europian life-form that looks like a huge clump of kelp seaweed and is attracted to the ship's lights. The lone, surviving astronaut, Professor Chang, uses his space suit radio to tell the American and Soviet astronauts, "There is life on Europa!" Europa's marine biosphere is modeled after deep-sea thermal vents on Earth: biomes where no light reaches, and energy for life originates from dissolved chemicals and settling organic detritus. Clarke describes "pipes and chimneys deposited by mineral brines gushing from the interior" of Europa, where "all the chemicals for life" are present. In this region, there are plants with spidery structures, "bizarre slugs and worms", crabs, and spiderlike creatures. Clarke also creates biomes along the warm banks of molten lava flowing deep underwater, where high pressure prevents water from evaporating into steam. He describes large kelplike growths that look like "banyan trees from Earth's tropics" and fishlike creatures without gills. Europa is an oxygenless world; all metabolism is sulfur-based. Despite Europa's biological diversity and abundance, Clarke points out that Europa is a doomed world and not destined for advanced evolution because its only energy source is internal thermal processes, and therefore very limited with respect to an evolutionary time scale. He solves this problem at the end of *2010: Odyssey Two* by having Jupiter ignite into a small star, providing Europa with a continuous source of heat and energy.

Clarke also provides a tour of Jupiter, starting in the upper atmosphere and going to the planet's core. His description of Jupiter's atmosphere is similar to that in *A Meeting with Medusa*; in fact, he uses much of the same language, for example, a "witch's cauldron of hydrochemicals" and "petro-chemicals", although he does mention this time that the petrochemicals are formed from lightning in the upper atmosphere. In reality, hydrocarbons are indeed present in the atmosphere. Small amounts of methane and even smaller amounts of acetylene and ethane have been detected in Jupiter's upper atmosphere (29). UV radiation from the sun breaks methane into radicals that re-form into acetylene and ethane. On the basis of radical chemistry, ethylene, benzene, and methylacetylene are probably also present. Further down in the atmosphere, ammonia and phosphine (PH_3) are detected (29). White clouds in this region consist of ammonia crystals, and the colored clouds are believed to consist of ammonium hydrogen sulfide, phosphorus compounds, and organic carbon compounds. Below the clouds, water vapor, carbon monoxide, and germane (GeH_4) are detected with IR spectroscopy (29). Beyond this, little is actually known about Jupiter's structure.

As the tour descends through the atmosphere, various life-forms are encountered: giant gas bags several kilometers in diameter grazing on "aerial mountains of hydrocarbon foam". Smaller organisms shaped like aircraft hover around acting either as predators or parasites of the giant gas bags.

After descending through several cloud layers of "petrochemicals" and superheated steam, Clarke describes an abrupt discontinuity of only a few kilometers that is "denser than any rock on Earth". He explains that it is composed of various layers of different and complex silicon and carbon compounds. Continuing further in Clarke's description, half-way to the core the pressures and temperatures are so great that all molecules are broken into their elemental constituents. Eventually a deep sea of hydrogen is reached that is liquid and metallic. At the core a solid surface is finally encountered, composed of a diamond the size of the Earth. Clarke bases the diamond core on a paper (which he actually references in *2010: Odyssey*

Two) by Marvin Ross that proposed that methane in giant gas planets will eventually sink to the core, and because of the high pressures and temperatures, crystallize to diamond (40). Ross proposed this scenario for Uranus and Neptune, and Clarke extended it to Jupiter.

In *2061: Odyssey Three*, the surface of Halley's Comet is explored by the ship *Universe*. Clarke portrays the surface as covered with black snow, geysers, and "rocky reefs embedded in water and hydrocarbon ice". The comet is full of unusual organic geochemical features such as a solid "lake", named Lake Tuonela, composed of heavy organic tars, oils, and pitch. Cores of the comet's surface reveal complex organic compounds, similar to frozen coal tars, but not of biological origin. A cave containing waxy hydrocarbon stalactites and stalagmites, in addition to iron pyrite, is also present on Halley's Comet.

Spectroscopic analysis of comet tails and the coma surrounding the comet nucleus indicate that they consist primarily of water, ammonia, and methane ice, in addition to small amounts of various metals such as sodium, potassium, calcium, iron, nickel, vanadium, manganese, cobalt, copper, and chromium (41). Sometimes silicates are present in comet tails as small particles (approximately one micrometer in diameter) giving rise to "dust-tails". In general, comets are essentially "dirty snowballs" or ice conglomerates and are either by-products of the solar system's origin or are interstellar material captured by the sun's gravitational field (41).

The composition of dust grains expelled by Halley's Comet was examined by the probes *Giotta* and *Vega* (1968) (29). Both traveled close enough to sample small dust grains using mass spectrometry (MS). The dust grains, upon impact with a metal target of silver or platinum, were vaporized and all molecules broken into the constituent elements. Likewise, ionization of the elements occurred as a result of the impact and molecular disintegration. In this manner, approximately 1500 grains were analyzed. On the basis of chemical composition, three types of dust grains were identified. The first contains C, O, Na, Mg, Si, Ca, Fe, and some S and Cl. These grains essentially consist of a silicate-type matrix. The second type contains H, C, N, and O and is thought to be composed of organic molecules rather than water, carbon dioxide, and hydrogen cyanide ices, which presumably would have evaporated in space before reaching the probe. The third grain type, and most abundant, is a composite of the first two types. Mayo Greenberg (29) demonstrated in the laboratory that such composite grains can be formed in interstellar space at very low temperatures. At temperatures of approximately 10 K, water, carbon dioxide, and hydrogen cyanide condense around silicate particles. Over several hundreds of millions of years, UV radiation causes molecular radicals of water, carbon dioxide, and hydrogen cyanide to form, and they in turn can re-form into larger, more complex organic molecules that are resistant to evaporation.

As always, Mars is still a popular planet to visit, as illustrated in *Mars* (42) by Ben Bova and *Red Mars* (43) and *Green Mars* (44), both written by Kim Stanley Robinson. Ben Bova's *Mars* chronicles the first manned mission to Mars. He accurately describes the atmosphere as mostly carbon dioxide and with such a low pressure that "[a]n unprotected human would die in an explosive agony of ruptured lungs, and blood … would literally boil at such low pressure." Compare this description of the Martian atmosphere to those provided by Weinbaum and Miller! The search for liquid water and life are the major quests undertaken in Bova's *Mars*. It is generally believed that where there is liquid water, a good possibility for life exists. Ice is first found in a soil coring of the permafrost beneath a rock containing a streak of copper oxide. At first, the green copper oxide is thought to be Martian lichen, especially when ice is discovered less than a meter below. The permafrost coring reveals that the water has a high concentration of dissolved carbon dioxide, in addition to minerals such as iron and silicates. Additional corings also reveal that the "soil is loaded with superoxides", which degrade into ozone. Bova has the scientists analyze the soil with a light-weight, portable gas chromatograph–mass spectrometer (GC–MS), which he describes as analyzing "the chemical composition of materials, virtually atom by atom." Though Bova's description of the GC–MS is correct, he mistakenly has a scientist "separate out the dissolved minerals" from collected permafrost water using centrifugation!

As the story progresses, the search for life appears to become futile as analysis of soil samples reveals only traces of organic compounds, but no life. One biologist proposes that organic compounds necessary for life should be present in the Martian soil, brought by chondritic meteorites bearing "amino acids and other long-chain carbon molecules", but is rebuffed by another biologist who proposes that the high UV radiation levels, in addition to the subfreezing temperatures, lack of liquid water, and the presence of superoxides in the soil, pretty much rule out any chance for Martian life. Only when they reach the canyon, Tithonium Chasma, do they discover "frail gray feathers of clouds … wafting through the vast canyon far below." Further down on the canyon floor, they find a higher heat flow from below ground than there was on the plain where they landed and started their explorations. The canyon floor is composed of a sandy regolith that contains liquid water rather than a permafrost, and some of the rocks are found to contain "orange intrusions" that the geologists speculate are "some sort of sulfur compound." Analysis reveals that these "orange intrusions" are a lichen similar to "the crustose thalli of Antarctica". This Martian lichen is described as having a "hard silicate shell to protect them from the cold, but the shell is water-permeable" with "windows in it that allow sunlight through." The biologists hypothesize that the windows are transparent to IR and visible light. They also discover that these lichen lace their internal water supply with "some form of alcohol" to act as a natural antifreeze.

Besides descriptions about water and life, Bova also speculates about Martian geology. Jamie, the mission's geologist, explores the cliffs of two different canyons. The walls of the first canyon, Noctis Labyrinthus, are composed of a single material; there is no stratification to indicate layering of various sediments and minerals by different geological processes. Jamie hypothesizes that the "reddish gray wall" is actually exposed mantle of iron that never sank to the core, hence, a possible reason for the large amount of iron oxide on the Martian surface. The second canyon, Tithonium Chasma, is stratified, and the upper layer under the caprock is soft, crumbly, and easily broken down. The second layer is deeper red, containing more iron, possibly shergottite. The third layer, infused with "yellowish intrusions", is proposed to be bauxite.

In addition to canyon walls, an old lava flow from the extinct volcano, Pavonis Mons, is analyzed. The scientists find that the lava is a dark basaltic rock, rich in iron. Attempting to determine its age, they analyze the potassium-to-argon ratio and determine it is approximately five million years old. A second analysis of the uranium-to-lead ratio reveals that the lava flow is much older; leading several of the geologists to propose that argon was outgassed, giving rise to incorrect results.

Both of these radioisotope elemental ratios are used in geochemistry as geochronometers (45). Their basis is the radioactive decay of a long-lived radionuclide into a stable daughter product. By measuring the relative amounts of parent and daughter products, in addition to knowing the decay constant, the age of a geological sample can be measured. This is, of course, assuming that the system remained closed, and there was no loss or gain of the parent or daughter nucleotide except from radioactive decay. The isotope ^{40}K decays into two daughter products: ^{40}Ca and ^{40}Ar. Measurement of ^{40}Ca is not useful because ^{40}Ca is also the most abundant isotope of calcium. Even though ^{40}Ar is the most abundant isotope of atmospheric argon, potassium-bearing minerals usually do not contain argon; therefore, the age of a sample can be determined by measuring the amount of ^{40}K and ^{40}Ar and using the decay constant of 5.54×10^{-10} per year. Likewise, a radioactive decay series exists between uranium, thorium, and lead. Use of this system as a geochronometer is a bit more complicated because ^{238}U decays to ^{206}Pb, ^{235}U to ^{207}Pb, and ^{232}Th to ^{208}Pb. Each of these parent–daughter systems provides an independent age, which in an ideal situation, should all agree.

The atmosphere, meteorology, and geology of Mars are also presented accurately in Kim Stanley Robinson's novels *Red Mars* (43) and *Green Mars* (44). In *Red Mars*, Robinson describes the travel to and exploration of Mars by a international team of 100 scientists. He presents in great detail and depth the technological challenges they must overcome as they begin to establish a base and conduct scientific research. *Red Mars* also describes the eventual colonization and resource exploration of Mars by large multinational corporations, and the resulting social, economic, and ecological crises that arise.

Robinson's description of the physical geology of the Martian surface is beautiful, stunning, and spectacular. In great detail and vividness, he describes the various terrains of Mars: deserts, polar caps, canyons, rocky plains, and mountain ranges, conjuring images so graphic that it is difficult to believe he has not actually been there. He does this again in a novella also entitled *Green Mars* (46) in which he describes a mountain-climbing expedition (in great detail with charts and maps!) of Olympus Mons, "the tallest mountain in the solar system".

One of the major themes of *Red Mars* and *Green Mars* is terraforming the planet into a viable environment for humans. Robinson brings up many technical and environmental problems encountered, and describes the numerous projects that scientists and engineers are using to transform Mars.

The first major problem in terraforming Mars is altering the atmosphere so that it is thicker, contains more oxygen and water vapor, and has greatly reduced carbon dioxide levels. One method to increase water vapor is to increase the temperature of the planetary surface causing the polar ice caps to melt. This undertaking is attempted in several ways: drilling of "mole holes" deep into the planet so that geothermal heat is released into the atmosphere, and specifically emitting carbon tetrafluoride, hexafluoroethane, sulfur hexafluoride, methane, and nitrous oxide, all of which are powerful greenhouse gases and absorb in the 8- to 12-micrometer range, a region where water and carbon dioxide do not absorb. Water vapor is also added to the Martian atmosphere by bringing in large "ice asteroids" from the asteroid belt and exploding them in the upper atmosphere. Their fragments vaporize upon falling toward the Martian surface. Another project developed in *Green Mars*, by a giant multinational corporation, is the building of a giant lens with several mirrors that is placed into Martian orbit. Using this device, the *Soletta*, sunlight is focused into a beam reaching 5000 °F and used to melt sand and regolith, releasing volatiles into the atmosphere. Many scientists are opposed to this procedure because, in addition to water vapor, CO_2 is a major component released, a gas that they are trying to reduce if the surface is ever to be viable for humans.

A second problem encountered with terraforming Mars is the low level of nitrogen. The "lack of nitrogen was in fact one of the biggest problems that the terraforming effort faced; they needed more than they had, both in the air and in their soil." Without nitrogen, plants introduced on the Martian surface have a very difficult time growing, even as the surface temperature increased to viable levels, the atmospheric pressure increased from 10 to 160 millibars, and the water moisture is high enough that free-flowing streams are present on sunny days. Nitrogen-fixing microbes are used to help plants grow, but usually to no avail because of the low N_2 atmospheric concentrations. Any nitrate deposits found on Mars are quickly mined and either used as fertilizer or released into the atmosphere as N_2.

Finally, a major problem with terraforming is determining what the final composition of the atmosphere should be, relative to the chemical material available and the desired final result. Sax Russell, one of the first hundred scientists to land on Mars, lists the ideal mixture for the atmosphere to be 300 millibars nitrogen; 160 millibars oxygen; 30 millibars argon, helium, etc; and 10 millibars carbon dioxide, providing a total pressure of 500 millibars.

> The total pressure had to be high enough to drive oxygen into the blood, and 500 millibars was what was obtained on Earth at about the 4000-meter elevation.... Given that it was near the oxygen limit, it would be best if such a thin atmosphere had more than the Terrain percentage of oxygen in it, but it could not be too much or else fires might be hard to extinguish. Meanwhile CO_2 had to be kept below 10 millibars, or else it would be poisonous. As for nitrogen, the more the better, in fact 780 millibars would be ideal, but the total nitrogen inventory on Mars was now estimated at less than 400 millibars, so 300 millibars was as much as one could reasonably ask to be put into the air.

The lack of atmospheric nitrogen on Mars gives rise to proposed projects such as harvesting nitrogen from the pure nitrogen atmosphere of

Titan, one of Saturn's moons, liquefying it, and sending it to Mars. More realistically, argon will be used instead of nitrogen as the inert atmospheric component, even though it will be difficult to release from the regolith. Other proposals for a terraformed Mars stopped at a CO_2-rich atmosphere, an atmosphere that was warm and thick enough to allow plant growth and only require an oxygen face mask for humans.

As the terraforming of Mars progresses, Robinson describes the atmospheric and geological changes that occur. He scientifically explains why the colors of the sky change during the terraforming process, as a result of the gaseous composition, the presence of fines in the upper atmosphere (i.e., very small particles that remain airborne for long periods and are not scrubbed out by clouds), and the atmospheric pressure. He even explains and applies Rayleigh's law of scattering in his calculations for the color of the sky. Geological changes incurred are mainly due to increased amounts of liquid water present on the Martian surface and the concurrent erosion. The ice melting from the glaciers on summer afternoons produce rivulets of running water "cutting the land into new primitive watersheds, and turning talus slopes into what ecologists called "fallfields", those rocky patches … after the ice receded, their living components made up of algae and lichens and moss." He also states that "Mars had become so arid that when water touched it there were powerful chemical reactions—lots of hydrogen peroxide released and salt recrystallizations."

As Mars continues to be terraformed, Robinson engages in greater speculation as to the changes and resulting environmental chemistry. In addition, he provides great detail regarding the various algal, lichen, moss, and plant species, genetically engineered and otherwise, that are slowly gaining a foothold in protected and warm regions of the surface. The genetically engineered traits of these plants are emphasized, such as salt-tolerance, the ability to withstand subfreezing temperatures, high carbon dioxide levels, and the paucity of nitrogen. Robinson continues his terraforming of Mars in his novel *Blue Mars*, published in 1996.

Conclusion

Throughout the history of science fiction, the planets have held a special place for fertile imaginations. In the early years, very little was known or understood about the planets, and lack of knowledge this was reflected by the strong emphasis on adventure or romance. As our knowledge of the planets increased, so did the technical and scientific details and information presented. But it was not until the exploration of the Moon, unmanned spacecraft explorations of the planets, and the landings on Venus and Mars that science fiction really began to discuss the planets realistically. Now, science fiction accurately deals with the scientific and technical challenges presented to future explorers of the planets. This kind of science fiction, thick with science and light on speculation, presenting only plausible conjecture, is inspiring to read. Undoubtedly, as we learn more about the planets, including our own Earth, science fiction will continue to mature and become more sophisticated.

References

1. *Science News*, **1995**, 148, 26, 27.
2. Astor, John Jacob. *A Journey in Other Worlds*; 1894.
3. Pope, Gustavus W. *Wonderful Adventures on Venus*; 1895.
4. Greg, Percey. *Across the Zodiac: The Story of a Wrecked Record*; 1880.
5. Cromie, Robert. *A Plunge into Space*; 1890.
6. Pope, Gustavus W. *A Journey to Mars*; 1894.
7. Lasswitz, Kurd. *Two Planets*; 1897.
8. Gratacap, Louis Pope. *The Certainty of a Future Life on Mars*; 1903.
9. Wicks, Mark. *To Mars via the Moon*; 1911.
10. Burroughs, Edgar Rice. *A Princess of Mars*; 1917.
11. Thomas, Jane Fredrick. *To Venus in Five Seconds*; 1897.
12. Serviss, Garrett P. *A Columbus of Space*; 1911.
13. Verne, Jules. *From the Earth to the Moon*; 1865.
14. Serviss, Garrett P. *The Moon Metal*; 1900.
15. Wells, H. G. *The First Men in the Moon*; 1901.
16. Barr, Robert. *The Doom of London*, 1892.
17. Weinbaum, Stanley G. *A Martian Odyssey*; 1934.
18. Moore, Patrick; Hunt, Garry. *Atlas of the Solar System*; Rand McNally and Company: New York, published in association with the Royal Astronomical Society, 1983.
19. Miller, Walter M., Jr., *Crucifixus Etiam*; 1953.
20. Clarke, Arthur C. *The Sands of Mars*; 1952.
21. Silverberg, Robert. *Sunrise on Mercury*; 1957.
22. *Tomorrow's Worlds*; Silverberg, Robert, Ed.; 1969.
23. Herbert, Nigel. *The Planets: Portraits of New Worlds*; Penguin Books: New York, 1994.
24. Lewis, J. S. *Icarus*, **1972**, 16, 241–252.
25. Goettel, K. A.; Barshay, S. S. "The Chemical Equilibrium Model for Condensation in the Solar Nebula: Assumptions, Implications, and Limitations", in *The Origin of the Solar System*, Dermott, S. F., Ed.; John Wiley and Sons: New York, 1978, pp 611–627.
26. Simak, Clifford D. *Desertion*; 1944.
27. Moore, Patrick. *The New Atlas of the Universe*; Mitchell Beazly Publishers, 1984.
28. Weinbaum, Stanley G. *The Planet of Doubt*; 1935.

29. *The Cambridge Atlas of Astronomy*, 3rd ed.; Auduze, Jean; Israel, Guy, Eds.; Cambridge University Press: New York, 1994.
30. Clarke, Arthur C. *Before Eden*; 1961.
31. Pohl, Frederik. *The Merchants of Venus*; 1972.
32. Clarke, Arthur C. *A Meeting with Medusa*; 1971.
33. Blish, James. *How Beautiful with Banners*; 1966.
34. Baston, Raymond M. *Voyager 1 and 2, Atlas of Six Saturnian Satellites*; NASA Scientific and Technical Information Branch: Washington, DC, NASA SP-474, 1984.
35. Panshin, Alexei. *One Sunday in Neptune*; 1969.
36. Niven, Larry. *Wait It Out*; 1968.
37. Clarke, Arthur C. *2010: Odyssey Two*; 1982.
38. Clarke, Arthur C. *2061: Odyssey Three*; 1987.
39. Clarke, Arthur C. *2001: A Space Odyssey*; 1968.
40. Ross, M. *Nature*, **1981**, *292*, 435–436.
41. *Comets*; Brandt, John C., Ed.; W. H. Freeman and Company: San Francisco, CA, 1981.
42. Bova, Ben. *Mars*; 1992.
43. Robinson, Kim Stanley. *Red Mars*; 1993.
44. Robinson, Kim Stanley. *Green Mars*; 1994.
45. Richardson, Steven M.; McSween, Harry Y., Jr. *Geochemistry: Pathways and Processes*; Prentice Hall: Englewood Cliffs, NJ, 1989.
46. Robinson, Kim Stanley. *Green Mars*; 1985.
47. *The Planets*; Preiss, Byron, Ed., Bantam Books: New York, 1985. (Besides scientific information, this book includes short science fiction stories by Harry Harrison, William K. Hartman, Ray Bradbury, Gregory Benford, Roger Zelazny, Philip Jose Farmer, Jack Williamson, Frank Herbert, Marta Randall, and Paul Preuss.)

Appendix: Physical Planetary Data

Planet	Mean Distance from Sun (km)	Diameter (km)	Surface Gravity (Earth = 1)	Volume (Earth = 1)	Mean Density (gm/cm³)
Mercury	57,910,000	4,878	0.377	0.056	5.43
Venus	108,200,000	12,104	0.902	0.86	5.24
Earth	149,597,900	12,756	1.000	1.000	5.52
Mars	227,500,000	6,794	0.379	0.15	3.93
Jupiter	778,360,000	142,800	2.69	1,323	1.32
Saturn	$1,427 \times 10^6$	120,000	1.19	752	0.70
Uranus	$2,870 \times 10^6$	51,800	0.93	64	1.25
Neptune	$4,497 \times 10^6$	49,500	1.22	54	1.77
Pluto	$5,900 \times 10^6$	2,400	0.20	0.01	4.7

⋆5⋆

Beryllium, Thiotimoline, and Pâté de Foie Gras

Chemistry in the Science Fiction of Isaac Asimov

Ben B. Chastain

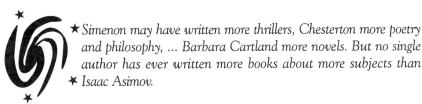

★*Simenon may have written more thrillers, Chesterton more poetry and philosophy, ... Barbara Cartland more novels. But no single author has ever written more books about more subjects than* ★ *Isaac Asimov.*

So wrote Peter Stoles in *Time* magazine a few years ago (*1*). Asimov has been called a "writing machine". The number of volumes that he wrote, co-wrote, or edited reaches nearly 500, of which more than 200 are currently in print. He is arguably the best-known science fiction writer in the world, and his "Foundation" series and "Robot" series introduced many to the joys of science fiction. However, most of his hundreds of books are nonfiction, including volumes on Shakespeare, the Bible, Gilbert and Sullivan, "Paradise Lost", and lecherous limericks, as well as many, many books explaining science to the nonspecialist. In 1965 he received the ACS Grady Award for science writing; one of his books still in print is *A Short History of Chemistry*.

Although he held a Ph.D. in chemistry from Columbia University, Asimov used chemistry as the driving force in only a few of his plots. This chapter offers a brief survey of what I consider to be important, or just characteristic, examples of chemistry in the stories of the best-known science fiction writer in the world.

Asimov's Life and Early Stories

It is frequently revealing to consider the fiction of an author in biographical context; with Isaac Asimov it is both enlightening and relatively easy, for his prodigious output includes a two-volume autobiography covering his first 58 years (2). A third volume was scheduled for the year 2000, but deteriorating health caused him to produce instead a "memoir", entitled *I. Asimov*, which was published by Doubleday in 1994.

Beginning at age 18 he kept a sort of diary or log containing some details of his daily life and thought along with a meticulous record of writings started, finished, submitted, rejected, accepted, published, and paid for. Thus we have available in his own words the when and the why of most of his stories. Obviously much of this chapter is based on material from these volumes.

Isaac Asimov was born in the Russian town of Petrovichi in 1920. His family emigrated to the United States in 1923. Beginning in 1926 his father owned a series of candy stores in Brooklyn; in these candy store newsstands Isaac first encountered the science fiction pulps of the 1930s. He had tried his hand at writing fiction as early as age 11, but did not attempt a real science fiction story until 1937, by which time he was a premed student at Columbia University. He had begun as a zoology major, but switched to

Biographical Summary	
1920	Born in Petrovichi, Russia
1923	Emigrated to the United States; settled in Brooklyn
1928	Naturalized as U.S. citizen
1935–1939	Columbia University (B.S., chemistry)
1939–1941	Columbia University (M.A., chemistry)
1942–1945	Chemist, Philadelphia Navy Yard
1945–1946	Private, U.S. Army
1946–1948	Columbia University (Ph.D., chemistry)
1949–1958	Professor, Boston University Medical School
1958–1992	Full-time writer

chemistry in his sophomore year because his conscience bothered him at having to kill a cat to dissect, and because he could not see anything in the microscopes in embryology.

That first story, called "Cosmic Corkscrew", does not exactly involve chemistry, but is worthy of mention because it sets the pattern for much of his fiction. He had read about the theoretically postulated, but as yet undetected, neutrino. Of course, the reason it was not detected then is that, lacking both charge and rest mass, it was beyond the scope of the available techniques. But, the 17-year-old Isaac speculated, what if the neutrino could not be detected because it flashed between past and future? It then might be used as a source for time travel. Quoting Asimov (3):

> That turned out to be typical of my science fiction. I usually thought of some scientific gimmick and built a story about that. … [A]s far as possible I was interested in realistic science, or the illusion of it. Even in that first story, I went to some trouble to explain about neutrinos as authentically as I could, for instance, even if I did introduce the time-travel angle out of left field.

In 1938 he finished "Cosmic Corkscrew" and carried it in person to the offices of *Astounding Science Fiction* magazine, where he met John W. Campbell, the then 28-year-old editor who was to become a legend in science fiction. The story was rejected, but a friendship was begun. During 1938 Asimov wrote several more stories, also rejected, including the first that might be said to be a chemistry story. It was called "The Weapon", and it was eventually published in 1942 in *Super Science Stories*. It can be found in the first volume of the autobiography. It is perhaps not a very good story, but for an 18-year-old just entering his senior year in college, it's not bad.

The chemistry and biology invoked in the story are obviously right out of his college courses; the stimulus for it was the gathering war clouds in Europe. The setting is the planet Mars, with a very humanlike race of Martians who have conquered the violent emotions of fear, anger, and hate, and so live in peace and harmony. An Earthman has come to beg them for the secret of their success. It turns out to lie in endocrinology; the Martians have discovered a way to stimulate and activate the pineal gland, that pea-sized organ in the brain for which no specific hormone has been isolated in humans. Its proper function, according to the story, is to serve as counterweight to the fear- and anger-producing adrenal glands. As the hero eventually discovers, the key is a specific short-lived isotope of bromine, needed by the pineal just as iodine is needed by the thyroid. He leaves Mars with the "weapon" needed to produce peace on Earth.

Another story from that pivotal year of 1938 was called "Ammonium" (though it was published in 1940 under the title "Magnificent Possession") (4). It is, on Asimov's own five-star scale, only a 1½-star story, but it is based on another bit from general chemistry. As early as the time of Lavoisier, it had been noted that ammonium compounds greatly resembled those of sodium, potassium, and other alkali metals: Asimov must have wondered what "ammonium metal" would be like, and how it might be obtained. The story presents a "complex derivative of hydrazine" called Ammonaline, which is able to abstract ammonium metal from an amalgam produced by substitution into a sodium amalgam.

$$Na + Hg \rightarrow Na\text{--}Hg \text{ (amalgam)}$$

$$Na\text{--}Hg + \text{"}NH_4Cl \text{ sol"} \rightarrow NH_4\text{--}Hg$$

$$NH_4\text{--}Hg + \text{"Ammonaline"} \rightarrow NH_4 \text{ (metal)}$$

$$NH_4 + \text{dry air} \rightarrow (NH_4)_2O$$

The dramatic thrust is that a pure, stable ammonium oxide is produced with all sorts of commercial potential; unfortunately, its indescribably beautiful appearance is accompanied by an indescribably horrible odor.

One of the early robot stories contained some coordination chemistry. "Runaround", published in *Astounding Science Fiction* in 1942 (5), deals with the use of robots to aid in mining the planet Mercury. One of the robots seems to be disobeying an order, something that should not be possible. It turns out that the robot is responding to a perceived threat to its own existence, namely, the presence of large amounts of carbon monoxide. At the temperature on Mercury's surface, the iron in the robot would react with the carbon monoxide to form the volatile iron carbonyl.

From these tentative beginnings came a steady maturation in writing style and a wellspring of ideas that had not dried up at the time of his death in April 1992. One of Asimov's stories in which chemistry is central to the plot came about because Robert Silverberg, at a 1971 science fiction convention, while arguing for the preeminence of human interest over science in science fiction stories, made an off-the-top-of-the-head remark about "some trivial matter concerning … plutonium-186." Asimov later kidded him about his nonexistent, impossible example; Silverberg shrugged it off as unimportant. Then Asimov said, "But just to show you what a real science fiction writer can do, I'll write a story about plutonium-186." Silverberg replied that he was putting together an anthology and would publish such a story if it met his "minimum standard of literacy".

His bluff called, Asimov began to work out the conditions under which such an impossible isotope might exist and the complications that would ensue. The resulting "story" became the first part of a three-part novel. The titles of the three parts constitute a quotation from Friedrich Schiller, "Against stupidity ... the gods themselves ... contend in vain." The novel is called *The Gods Themselves* (6). The plot involves alternate parallel universes (a common science fiction theme) in which the laws of nature may differ slightly.

In one such universe, the strong nuclear force (which holds the protons together in the nucleus) is even stronger than it is in our own, and therefore fewer neutrons are needed to produce a stable nucleus. The other universe establishes contact and changes a sample of stable tungsten-186 into highly unstable plutonium-186 by removing 20 electrons:

$$^{186}_{74}W \rightarrow {}^{186}_{94}Pu + 20 \, _{-1}e$$

The resulting isotope is a positron emitter, and so a source of much energy. In the other universe the electrons are used to change stable plutonium-186 into unstable tungsten-186, which is an energy source there. With a net transfer of 20 electrons from our universe into theirs, both universes experience a large energy gain. The result is called the Inter-Universe Electron Pump, and can produce low-cost energy for billions of years. The complications make the story, but the nuclear chemistry has explained plutonium-186.

Beryllium

Another story in which chemistry is central to the plot, intended for a 1953 anthology that never came out, was called "Sucker Bait" and was published in *Astounding Science Fiction* in 1954 (7). It concerned a very Earthlike planet whose attempted colonization by humans had resulted in the deaths of the colonists from an unexplained respiratory ailment. The major plot gimmick is the existence of an organization called the Mnemonic Service, whose men and women are trained to remember everything they see, hear, and read so that they can make connections—a sort of "natural" artificial intelligence, if you will. The worlds are filled with computers filled with data so that answers to anything are available if one knows what questions to ask. But the computers have no imagination. So humans are needed to supply the intelligence and imagination. These persons are seen as weird by most other humans and are feared and shunned.

The mystery in this story is solved by a Mnemonic who learns that the crust of the planet is unusually high in the lighter elements, including beryllium. He remembers from having read a very old book what humankind in general has forgotten—that beryllium is poisonous, that inhaling it in dust inhibits many enzyme systems that use other divalent metal ions, and that its toxic effects are usually seen in respiratory ailments. On other planets it had not mattered because the beryllium content of the soil was so low. In the distant past on Earth when people had used beryllium—in the first atomic piles and the first fluorescent lights—the toxicity was discovered and substitutes were sought. After they were found, people had no further use for beryllium, and after a long period during which it was not used, its toxicity was forgotten. So when the unusual beryllium-rich planet was found, its "sucker bait" was taken by people, and it killed them.

Thiotimoline

You will have noted, of course, that the chapter title contains an element, a compound, and a mixture. We have dealt with the element, and now we turn to the compound thiotimoline. The background is as follows: In 1947 Isaac Asimov was a graduate student in chemistry at Columbia, having worked for the Navy and served in the Army during the war period. He was doing his doctoral research under the direction of Charles Dawson. The project involved enzyme kinetics, using a stopwatch as timer and the iodine–starch complex as indicator. The substrate was catechol (*ortho*-dihydroxybenzene), and as Asimov recounted (8): "Catechol, as it happens, is very readily soluble, especially when it exists as fluffy crystals that present a large surface to the water. The result is that as soon as catechol touches the surface of the water it dissolves. It just seems to vanish without ever penetrating the water's skin." And as he watched one morning, the thought came, if it were any more soluble, it would dissolve *before* it hit the water. There was a possible story. Or, better yet: the time was nigh when he would have to write up his research results in a dissertation, in the approved turgid, convoluted style. Why not write a sort of mock dissertation about a substance that dissolved before it hit the water? And so thiotimoline was born.

The paper appeared in the March 1948 issue of *Astounding Science Fiction* (9), and is reprinted in this volume in Chemists at Play. It reported that the structure of the compound, extracted from a shrub, was still unknown, but it contained many hydrophilic groups (14 hydroxys, two aminos, and one sulfonic acid), which would enhance its solubility. There was no diagram of the endochronometer used to measure time of solution (it had been reported in a previous communication), but its description was given:

> ... a cell 2 cubic centimeters in size into which a desired weight of thiotimoline is placed, making certain that a small hollow extension at the bottom of the solution cell—1 millimeter in internal diameter—is filled. To the cell is attached an automatic pressure micropipet containing a specific volume of the solvent concerned. Five seconds after the circuit is closed, this solvent is automatically delivered into the cell containing the thiotimoline. During the time of action, a ray of light is focused upon the small cell-extension ... and at the instant of solution the transmission of this light will no longer be impeded by the presence of solid thiotimoline. Both the instant of solution— ... recorded by a photoelectric device—and the instant of solvent addition can be determined with an accuracy of better than 0.01%. ... The entire process is conducted in a thermostat maintained at 25.00 °C to an accuracy of 0.01 °C.

The results may be summarized quickly. Table I in the original paper demonstrates that careful purification of the material is necessary. Figure 1 shows the plateau effect; that is, for a fixed mass of thiotimoline the negative solution time increases as the volume of solvent injected increases, but only up to a point; then it levels off. The plateau height (PH) in minus seconds and the plateau volume (PV) in milliliters are characteristic of the nature of ions present in the solvent and their concentrations.

Table I. Purification stages of thiotimoline.

Purification Stage	Average "T" (12 observations)	"T" extremes	% error
As isolated	–0.72	–0.25; –1.01	34.1
First recrystallization	–0.95	–0.84; –1.09	9.8
Second recrystallization	–1.05	–0.99; –1.10	4.0
Third recrystallization	–1.11	–1.08; –1.13	1.8
Fourth recrystallization	–1.12	–1.10; –1.13	1.7
First resublimation	–1.12	–1.11; –1.13	0.9
Second resublimation	–1.122	–1.12; –1.12	0.7

Note: "T" is time of solution in seconds.

Source: Reproduced with permission from *Astounding Science Fiction*, March 1948.

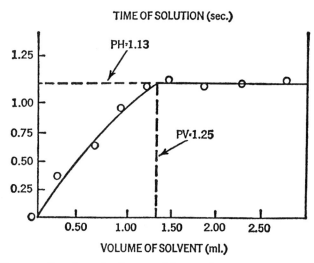

Figure 1. Plateau effect for a fixed mass of thiotimoline.

Source: Reprinted with permission from Astounding Science Fiction, March 1948.

Figure 2 in the original paper shows how PV changes with concentration of NaCl; this rate of change could be a rapid method for quantitative analysis of such solutions. Figure 3 shows how PH varies with the nature of the solute (constant ionic strength). Table II summarizes these effects; it shows that PV varies with ionic strength. Figure 4 suggests that mixtures of salts produce characteristic slope changes. This behavior means that endochronic measurements could be used for qualitative inorganic analysis as well as quantitative.

We have now covered the main chemical aspects of thiotimoline, although a second paper (in the December 1953 Astounding Science Fiction) (10) does include a description of a new apparatus used for purification by endochronic filtration. It involves two stopcocks, one of Y-shape, which are electrically controlled. When the circuit is closed, a vacuum pump starts, and exactly 5 seconds later a timer operates the stopcocks simultaneously. The result is that at about 0.72 second before the stopcocks open, the thiotimoline in the impure extract dissolves in the water that is about to be poured on it, and the solution is sucked out through a filter into a container. When the stopcocks do open, any impurities that dissolve in the water are carried into another container. The very pure thiotimoline thus produced has a PH of −1.124 seconds, slightly higher than the best previous result.

This second paper is entitled "The Micropsychiatric Applications of Thiotimoline" and discusses the possibility of withdrawing the water after

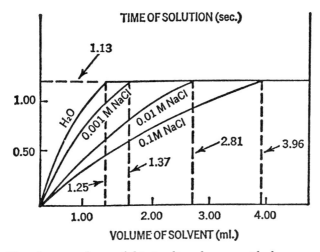

Figure 2. The plateau volume of thiotimoline changes with the concentration of NaCl.

Source: Reprinted with permission from *Astounding Science Fiction*, March 1948.

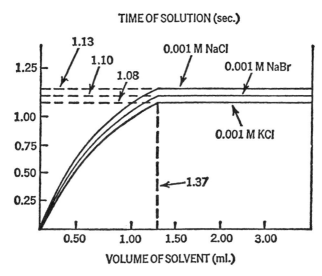

Figure 3. Plateau height varies with ionic strength.

Source: Reprinted with permission from *Astounding Science Fiction*, March 1948.

Table II. Summary of the Results of Figures 1–4 and Table I,
Showing That Plateau Volume Varies with Ionic Strength .

Solvent (Salt Solutions in 0.001 M Concentration)	Plateau Height (PH) (seconds)	Plateau Volume (PV) (milliliters)
Water	–1.13	1.25
Sodium chloride solution	–1.13	1.37
Sodium bromide solution	–1.10	1.37
Potassium chloride solution	–1.08	1.37
Sodium sulphate solution	–0.72	1.59
Calcium chloride solution	–0.96	1.59
Magnesium chloride solution	–0.85	1.59
Calcium sulphate solution	–0.61	1.72
Sodium phosphate solution	–0.31	1.97
Ferric chloride solution	–0.29	1.99

Source: Reproduced with permission from *Astounding Science Fiction*, March 1948.

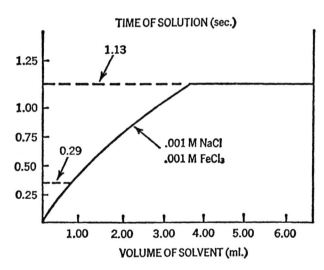

Figure 4. In a mixture of salts, each salt produces changes in its own characteristic slope.

Source: Reprinted with permission from *Astounding Science Fiction*, March 1948.

the thiotimoline has dissolved but before the water has actually been added, thus "fooling" it. The paper mentions a previous publication that has demonstrated the thermodynamic impossibility of this, but the idea leads to the science of "willometry", the measurement of human will by having the water added by hand and measuring the negative time of solution. Extremely strong-willed people achieved the full time produced by mechanical means; those with any hesitancy, even unconscious, had lesser effect. Persons with "split personalities" produced a situation in which at a given instant some of the thiotimoline had dissolved, but some had not. And so on.

A third paper in the series, a speech made to the American Chronochemical Society in 1960 (*11*), described the "telechronic battery" made by placing a number of endochronometers in series so that the solution of thiotimoline in one activated the water-delivering pipet in the next. Therefore, the thiotimoline in the second would dissolve 1.12 seconds before its water was delivered, which is 2.24 seconds before the water is delivered to the first. About 77,000 units in such a series would produce a final sample of thiotimoline that would dissolve a full day before the water was added to the first cell. The results of such a device include foretelling the future (if a certain horse wins a certain race tomorrow I will add water to the first cell, otherwise not; if my will is strong enough, I can find out a day ahead whether the horse will win) and even "endochronic bombs", which produce floods and hurricanes but leave no indication of having influenced the weather. Obviously we have gone far beyond chemistry.

A fourth and final round for thiotimoline came in a 1971 story "Thiotimoline to the Stars" (*12*), which described how thiotimoline-containing plastics and resins had been developed and bonded to metals producing large endochronic objects (like spaceships, for instance).

Pâté de Foie Gras

The final story to be considered here involves the mixture. "Pâté de Foie Gras" was first published in the September 1956 *Astounding Science Fiction* (*13*) and is also reprinted in this volume in Chemists at Play. It deals with nothing less than a goose that lays golden eggs! Or, to be precise, eggs that contain underneath the normal shell an inner shell about 2 millimeters thick made of pure gold. The yolk turns out to be heavy-metal poisoned with chloroaurate ion (about 3.2 parts per trillion). Studies produce the following analysis: "The chloroaurate ion is secreted by the liver into the bloodstream. The ovaries act as a trap for the ion, which is there reduced to metallic gold and deposited as a shell about the developing egg. ... The Goose finds this process useful as a means of getting rid of the gold atoms which, if allowed to accumulate, would undoubtedly poison it."

Where does the gold come from? Further tests find that a portion of the "heme" units in the blood are in fact "aureme" units; that is, they contain gold rather than iron. In fact, iron-56, the most common isotope, is missing altogether. The goose's liver (foie gras) turns out to be a living nuclear reactor, turning oxygen-18 into gold-197 through a series of steps in which iron-56 is an important intermediate. It also absorbs gamma rays harmlessly, and turns unstable isotopes into stable ones. The implications for radioactive waste disposal are staggering. But, of course, its eggs will not hatch. They are poisoned. Gold poisoned. No solution to the dilemma is offered. Instead, the readers are asked for their ideas. Even this fantasy follows the Asimov pattern. That is, except for the central assumption that nuclear reactions on a large scale can occur in living tissue, the biochemistry in the story is quite reasonable.

The Old Is the Basis

I close with a quotation from yet another Asimov book, not science or science fiction, but a murder mystery (involving academic chemists) (14). I think it speaks to everyone involved in chemistry. The subject is a history of organic chemistry being written by a retired professor. Another professor comments, "Oh, yes. Chemists need that book. Yes. Too many chemists live in the present only. Mathematicians and physicists know the history of their science because new developments supplement the old. In chemistry new developments seem to replace the old. Tendency is to forget the old, then; and too much is forgotten in that way. The old is the basis for the new. New can't be understood properly without the old."

"Quite right, murmured Brade." And so say I.

References

1. *Time*, February 26, 1979; p 79.
2. Asimov, Isaac. *In Memory Yet Green*; Doubleday: New York, 1979; Avon: New York, 1980. *In Joy Still Felt*; Doubleday: New York, 1980; Avon: New York, 1981.
3. *In Memory Yet Green*; p 171.

4. Asimov, Isaac. "Ammonium". In *Future Fiction* July 1940 (as "Magnificent Possession"), available in *The Early Asimov*; Doubleday: New York, 1972.

5. Asimov, Isaac. "Runaround". In *Astounding Science Fiction*, March 1942, available in *I, Robot*; Doubleday: New York, 1950.

6. Asimov, Isaac. *The Gods Themselves*. In *Galaxy*, March and May 1972; and *If*, April; available as book, Doubleday: New York, 1972.

7. Asimov, Isaac. "Sucker Bait". In *Astounding Science Fiction* February and March 1954, available in *The Martian Way and Other Stories*; Doubleday: New York, 1955.

8. Reference 3, p 497.

9. Asimov, Isaac. "The Endochronic Properties of Resublimated Thiotimoline". In *Astounding Science Fiction* March 1948, available in *The Early Asimov*; Doubleday: New York, 1972.

10. Asimov, Isaac. "Micropsychiatric Applications of Thiotimoline". In *Astounding Science Fiction* December 1953, available in *Only a Trillion*; Abelard-Schuman: New York, 1957.

11. Asimov, Isaac. "Thiotimoline and the Space Age". In *Astounding Science Fiction* October 1960, available in *Opus 100*; Houghton Mifflin: New York, 1969.

12. Asimov, Isaac. "Thiotimoline to the Stars". In *Astounding*; Random House: New York, 1973; Harry Harrison, Ed.; available in *Buy Jupiter and Other Stories*; Doubleday: New York, 1975.

13. Asimov, Isaac. "Pâté de Foie Gras". In *Astounding Science Fiction* September 1956, available in *Only a Trillion*; Abelard-Schuman: New York, 1957, also in *Where Do We Go From Here?* Asimov, I., Ed.; Doubleday: New York, 1971.

14. Asimov, Isaac. *A Whiff of Death*; Fawcett Crest: New York, 1978; p 187.

★6★

The Improbable Properties of Imipolex G

Chemistry and Materials Science in Thomas Pynchon's *Gravity's Rainbow*

Carl Trindle

★*Gravity's Rainbow* (1) is a monstrous work, in bulk and difficulty. A forbidding fat brick of a book, it has been praised as this half-century's reply to James Joyce (2). (This is already a warning.) It is a cult favorite, particularly among scientists. In fact I would never have persisted in reading this doorstop of a book, were it not for the extravagant praise given it by a noted physicist and interpreter of science, James Trefil. Here I thank him, in a way.

It may be a mistake to enter this book with the expectation of the virtues of other fat novels—compelling characters, a plot that grabs you by the lapels and forces you to keep reading, a sense of ending with all the loose ends tied neatly at the conclusion. It is closer to a long nightmare, in which association replaces logical connection, but nothing is purely chance. It is surely a world all its own, with occasional connections with the one we inhabit.

Gravity's Rainbow is a stew of themes and images. If writers are divided into those who take out and those who put in, Pynchon is a pack rat with ambition. Pynchon knows, and uses, imagery from popular culture: We become acquainted with classic films such as *Frankenstein* and *Metropolis*, the comedy of Abbott and Costello, radio and TV shows of all kinds and formats, Plastic Man and other comic book characters, and Rocketman and other superheroes. He finds sex and scatology good dirty fun. As we bump along in his carnival ride, we are told much of World War II, the time in which much of the story is set; Pavlovian psychology; rocket science; statistics and other applied math; and, inescapably, chemistry.

Pynchon's Influences

Thomas Pynchon's writing is often compared with Joseph Conrad's *Heart of Darkness*, but we can look to his own time for similar work. The school called "Cold War Gothic", which can include Norman Mailer, Tom Stoppard, J. D. Salinger, and Richard Fariña, would also welcome Pynchon. (It is no accident that Pynchon dedicated *Gravity's Rainbow* to Fariña.)

One critic suggested that Pynchon be considered "a gothic writer of the science fiction persuasion". The term "gothic" is to be taken in T. S. Eliot's sense, as writing describing "uncanny events, evoking terror by mystery, cruelty, and horror ... evil passionately felt. ...The novel makes use of the only gothic locale that retains any mystery and terror for us in a thoroughly secular, disenchanted age: the laboratory" (3).

Gravity's Rainbow is deeply hostile to science, despite skillful use of scientific images and styles. So, in what way does it appeal to certain scientists? I suggest, with no particular confidence, that there is some degree of entertainment value in the satirical use of the flat leaden prose of technical reports. Or could it be that as Samuel Johnson said in a chauvinistic moment, it is "like seeing a dog walk on its hind legs; it is not done well, but the wonder is that it is done at all"? Or, could it be that we scientists, who "have known sin", need literature's catharsis of nagging doubts and fears?

Who Is Laszlo Jamf?

No Gothic science fiction novel would be complete without an evil genius, and we soon are given to know something of Pynchon's character Professor-Doctor Laszlo Jamf. In an advertising brochure distributed by Agfa Company of Berlin in 1934, Jamf states (*1*, p 71, line 11):

> "Kryptosam" is a proprietary form of stabilized tyrosine, developed by IG Farben as part of a research contract with OKW. An activating agent is

included which, in the presence of some component of the seminal fluid to date [1934] unidentified, promotes conversion of the tyrosine into melanin, or skin pigment. In the absence of seminal fluid, the "Kryptosam" remains invisible. No other known reagent, among those available to operatives in the field, will alter "Kryptosam" to visible melanin. It is suggested, in cryptographic applications, that a proper stimulus be included with the message which will reliably produce tumescence and ejaculation. A thorough knowledge of the addressee's psychosexual profile would seem of invaluable aid.

There is a connection, (of course; there's nothing *but* connections) between Jamf and the "hero" of the book, Tyrone Slothrop. Slothrop's education was financed when, as an infant, he was the subject of conditioning experiments by that evil man. Tiny Tyrone was "taught" in the Pavlovian way to respond to the presence of a certain chemical by an involuntary physical response; that is, to associate his own infant erection with the presence of Imipolex G. Now the sites of arrival of the V-2 rockets to London can be mapped with Poisson precision on to the sites where Slothrop practices his appeal to women. Needless to say, Slothrop sets off to find the infamous Jamf as soon as he himself realizes the connection.

We learn something of Jamf's later work from Wimpe, the expert on cyclized benzylisoquinolines, who tells (*1*, p 348, line 7ff) the history of

Oneirine, and Methoneirine. Variations reported by Laszlo Jamf in the ACS Journal, year before last. Jamf was on loan again, this time as a chemist to the Americans, whose National Research Council had begun a massive program to explore the morphine molecule and its possibilities—a Ten-Year Plan, coinciding, most oddly, with the classic study of large molecules being carried on by Carothers of du Pont, the Great Synthesist. Connection? Of course there's one. But we don't talk about it. NRC is synthesizing new molecules every day, most of them from pieces of the morphine molecule. Du Pont is stringing together groups such as amides into long chains. The two programs seem to be complementary, don't they? The American vice of modular repetition, combined with what is perhaps our basic search: to find something that can kill intense pain without causing addiction.

Results have not been encouraging. We seem [to be] up against a dilemma built into Nature, much like the Heisenberg situation. There is nearly complete parallelism between analgesia and addiction. The more pain it takes away, the more we desire it. It appears we can't have one property without the other, any more than a particle physicist can specify position without suffering an uncertainty as to the particle's velocity.

But chemists are not swayed by the hint that there is some knowledge beyond our grasp, and that the bargain we strike to gain our understanding might be to our disadvantage, in the long run. After all, we were given insight in a most mystical, magical way, in a dream.

Pökler's Fantasia on Kekulé's Dream

The character Pökler relates (*1*, p 410, line 32),

> Look at this. There is about to be expedited, for Friedrich August Kekulé von Stradonitz, his dream of 1865, the great Dream that revolutionized chemistry and made the IG* possible. So that the right material may find its way to the right dreamer, everyone, everything involved must be exactly in place in the pattern. It was nice of Jung to give us the idea of an ancestral pool in which everybody shares the same dream material. But how is it that we are each visited as individuals, each by exactly and only what he needs? Doesn't that imply a switching-path of some kind? a bureaucracy?
>
> ...
>
> Here, here's the rundown on Kekulé's problem. Started out to become an architect, turned out instead to be one of the Atlantes of chemistry, most of the organic wing of that useful edifice bearing down on top of his head forever—not just under the aspect of IG, but of World, assuming that's a distinction you observe, heh, heh. ... Once again it was the influence of Liebig, the great professor of chemistry on whose name-street in Munich Pökler lived while he attended the T.H.† Liebig was at the University of Giessen when Kekulé entered as a student. He inspired the young man to change his field. So Kekulé brought the mind's eye of an architect over into chemistry. It was a critical switch. Liebig himself

*IG is a short term for IG Farben, the German chemical manufacturing company.
†T.H. is *Technische Hochschule*.

seemed to have occupied the role of a gate, or sorting-demon such as his younger contemporary Clerk Maxwell once proposed, helping to concentrate energy into one favored room of the Creation at the expense of everything else.

…

Young ex-architect Kekulé went looking among the molecules of the time for the hidden shapes he knew were there. … He *saw* the four bonds of carbon, lying in a tetrahedron—he *showed* how carbon atoms could link up, one to another, into long chains. … But he was stumped when it came to benzene. He knew there were six carbon atoms with a hydrogen attached to each one—but he could not see the shape. Not until the dream: until he was made to see it, so that others might be seduced by its physical beauty, and begin to think of it as a blueprint, a basis for new compounds, new arrangements, so that there would be a field of aromatic chemistry, to ally itself with secular power, and find new methods of synthesis, so there would be a German dye industry to become the IG. …

Kekulé dreams the Great Serpent holding its own tail in its mouth, the dreaming Serpent which surrounds the World. But the meanness, the cynicism with which this dream is to be used! The Serpent that announces, "The World is a closed thing, cyclical, resonant, eternally-returning," is to be delivered into a system whose only aim is to *violate* the Cycle.

…

No: what the Serpent means is—how's this—that the six carbon atoms of benzene are in fact curled around into a closed ring, *just like that snake with its tail in its mouth*, GET IT? "The aromatic ring we know today," [says] Pökler's old prof, Laszlo Jamf … "but *who*, … who sent, the *Dream?*" It is never clear how rhetorical any of Jamf's questions are. "Who sent this new serpent to our ruinous garden, already too fouled, too crowded to qualify as any locus of innocence … we had been given certain molecules, certain combinations and not others … we used what we found in Nature, unquestioning, shamefully perhaps—but the Serpent whispered, '*they can be changed*,' and new molecules assembled from debris of the given. … 'Can anyone tell me what else he whispered to us? Come—who knows? You. Tell me, *Pökler* …'"

His name fell on him like a thunderclap, and of course it wasn't Prof.-Dr. Jamf at all, but a colleague down the hall who had pulled reveille duty that morning. ...

What a word-drunk riff, a dream of a dream, on Kekulé's shaggy story, which has occasioned so much strife in the otherwise decorous Division of the History of Chemistry!

The Invention of Imipolex G

With the sense of power given in the dream, a chemist—or, anyway, *this* chemist, Jamf,—could hardly resist playing the sorcerer's apprentice. (*1*, p 249ff)

The origins of Imipolex G are traceable back to early research done at du Pont. Plasticity has its grand tradition and main stream, which happens to flow by way of du Pont and their famous employee Carothers, known as the Great Synthesist. His classic study of large molecules ... [was] an announcement of Plasticity's central canon: that chemists were no longer to be at the mercy of Nature. They could decide now what properties they wanted a molecule to have, and then go ahead and build it. At du Pont, the next step after nylon was to introduce aromatic rings into the polyamide chain. Pretty soon a whole family of "aromatic polymers" had arisen: aromatic polyamides, polycarbonates, polyethers, polysulfanes. The target property most often seemed to be strength—first among Plasticity's virtuous triad of Strength, Stability, and Whiteness (*Kraft, Standfestigkeit, Weiße*): how often these were taken for Nazi graffiti. ...L. Jamf, among others, then proposed, logically, dialectically, taking the parental polyamide sections of the new chain, and looping *them* around into rings, too; giant ... [macrocyclic] rings, to alternate with the aromatic rings. This principle was easily extended to other precursor molecules. A desired monomer of high molecular weight could be synthesized to order, bent into its ...[macrocyclic] ring, clasped, and strung in a chain along with the more "natural" benzene or aromatic rings. Such chains would be known as "aromatic" ... polymers. One hypothetical chain that Jamf came up with, right before the war, was later modified into Imipolex G.

Jamf at the time was working for a Swiss outfit called Psychochemie AG, originally known as the Grössli Chemical Corporation, a spinoff from Sandoz (where, as every schoolchild knows, the legendary Dr. Hofmann made his important discovery) [namely, LSD].

Trying to make chemical sense of this passage, I replaced the original word "heterocyclic" with "macrocyclic".

Schweitar's Testimony

Of course, there is a long way to go between a structural modification that seems reasonable in principle, and the actual development of the new form. Mario Schweitar, a consultant (spy) for Psychochemie AG, replies to Slothrop's inquiry (*1*, p 261, line 3):

"Gaaah—"

"Pardon me?"

"That stuff. Forget it. It's not even our line. You ever try to develop a polymer when there's nothing but indole people around? … Imipolex G is the company albatross, Yank. They have vice-presidents whose only job is to observe the ritual of going out every Sunday to spit on old Jamf's grave. You haven't spent much time with the indole crowd. They're very elitist. They see themselves at the end of a long European dialectic, generations of blighted grain, ergotism, witches on broomsticks, community orgies, cantons lost up there in folds of mountain that haven't known an unhallucinated day in the last 500 years—keepers of a tradition, aristocrats."

"*Wait* a minute … Jamf dead? You say Jamf's *grave*, now?" It ought to be making more of a difference to him, except the man was never really alive so how can he be really—

On the other hand, Jamf's persistence did pay (*1*, p 249).

Imipolex G has proved to be nothing more—or less—sinister than a new plastic, an aromatic …[macrocyclic] polymer, developed in 1939, years before its time, by one L. Jamf for IG Farben. It is stable at high temperatures, like up to 900 °C, it combines good strength with a low power-loss factor. Structurally, it's a stiffened chain of aromatic rings, hexagons like the gold one that slides and taps above Hilary Bounce's navel, alternating here and there with what are known as …[macrocyclic] rings.

What Can Jamf Mean?

Linked macrocycles were imagined as long ago as 1912 by Richard Willstät-ter, according to lore (4), and were first synthesized in the 1960s (5), just about the time Pynchon was writing all this! The difficulty of course is to meet the costs that must be paid to bring the free-floating ends of the chains together. The first methods produced little more than statistical yields (Figure 1).

Frisch and Wasserman (6) and Schill (7) visualized a ladder-formed molecule; if the ends were twisted before cyclization, and the rungs are then broken, a catenane such as the thyme ladder can result (Figure 2). The idea was proved practicable by Walba et al. (8). However, the fundamental sta-tistical barrier to the desired twisting was a serious drawback.

Recent advances in the synthesis of catenanes incorporate direct con-straints on the motion of the ends. The Schill–Lüttringhaus method (9), reported in 1964 (!), relies on constraints that are consequences of the pre-ferred short-range geometry of the species shown (Figure 3).

Already in the 1960s, that imaginative and fertile time, Wasserman envisioned template-directed synthesis using a transition metal as an attractant and organizer. In the 1980s Dietrich-Buchecker and Sauvage (10) developed a variety of clever and efficient means for synthesis of [2]catenands. (The notation means that two macrocyclic rings are linked topologically.) The limit to practical synthesis at present seems to be the [3]catenands (11), though the yield is excellent.

It may be forgivable, especially here, to mention some speculations dat-ing from the 1950s (!) on the properties of [n]catenands. Frisch, Martin, and Mark (12) described their physical studies of polysiloxene, some results of which they could not explain if they viewed the polymers as simple chains. Polysiloxene is ordinarily a viscous liquid, although polymers of simi-lar mass containing carbon or nitrogen are solids. Extensive cross-linking is required to make an insoluble film of this material. A polysiloxene that according to well-established rules of thumb should display an end-to-end distance in solution of 840 Å (as determined by light-scattering measure-ments) displays an anomalous value of 1600 Å, with a dissymmetry factor far beyond that expected for randomly coiled linear chains. A chain of like

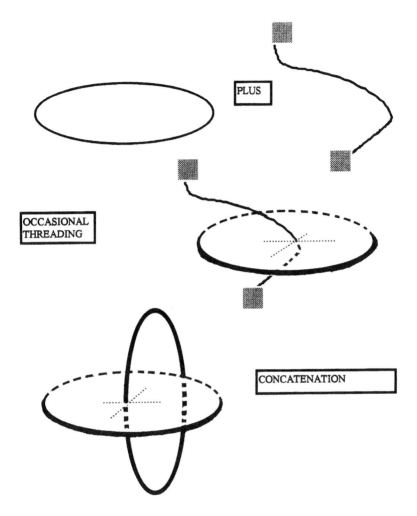

Figure 1. A ring and a chain can conceivably form a catenane if the chain threads its way through the ring, and then can be induced to close.

molecular mass, composed of 180 intertwined rings, each of which comprises about 50 monomers, would have an end-to-end distance of about 3000 Å, and would also display an unusual dissymmetry.

Patat and Derst (13), describing their kinetic studies of polymerization and depolymerization of phosphoryl nitrile chloride, $[-PCl_2N-]_n$ suggested that an insoluble, unreactive residue is composed of statistically interlinked

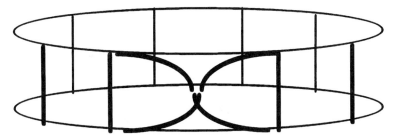

*Figure 2. Two open rings are connected by "rungs" of what will be a Thyme lad-
der. The rungs are represented by the vertical lines. It is possible that a crossover
joining the upper and lower rings can be effected; this crossover is represented by
the heavy curved lines in the foreground. Once the rungs are broken, the resulting
structure consists of two linked rings. The preliminary ordering makes this process
much more likely than the random threading shown in Figure 1.*

rings of about 10 to 50 trimers that make chains of molar mass near one
million. These would be semirigid, black materials of great durability—like
molecular chain mail.

Further Properties of Imipolex G

Pynchon's invention has still more remarkable properties (*1*, p 698ff):

> Imipolex G is the first plastic that is actually *erectile*. Under suitable stim-
> uli, the chains grow cross-links, which stiffen the molecule and increase
> intermolecular attraction so that this Peculiar Polymer runs far outside the
> known phase diagrams, from limp rubbery amorphous to amazing perfect
> tessellation, hardness, brilliant transparency, high resistance to tempera-
> ture, weather, shock of any kind.
>
> …
>
> Evidently the stimulus would have had to be electronic. Alternatives for
> signaling to the plastic surface were limited:
>
> (a) a thin matrix of wires, forming a rather close-set coordinate system
> over the Imipolectic Surface, whereby erectile and other commands could
> be sent to an area quite specific, say on the order of ½ cm²,

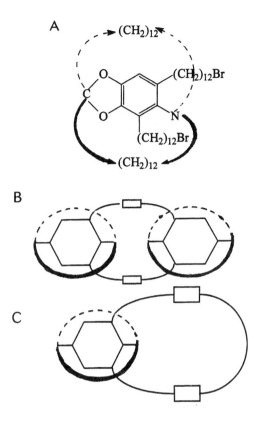

Figure 3. Schill and Lüttringhaus constructed an ordered precursor, shown as a chemical structure in part A. This structure can be schematized to the essential structures: a peripheral ring, an imbedded benzene ring, and two long reactive arms. If the precursor dimerizes, catenane B is formed. If the two arms are joined by a chain, catenane structure C results.

(b) a beam-scanning system—or several—analogous to the well-known video electron stream, modulated with grids and deflection plates …

(c) alternatively, the projection, *onto* the Surface, of an electronic "image", analogous to a motion picture. This would require a minimum of three projectors, and perhaps more. Exactly how many is shrouded in another order of uncertainty: the Otyiyumbu Indeterminacy Relation …

The cross-links must apparently involve labile electrons, which in the black flexible phase absorb light throughout the visible spectrum, leaving only the carbonyl and amide functionalities absorbing ultraviolet light. The drastic change in conformation and in bonding must be triggered by an electric field of modest strength. Do we know anything like that? Yes, already in 1961, Hardin McConnell (14) had set the stage for design of molecular switching devices, which was realized as "Ratner's diode" in 1974 and as a molecular shuttle by Farazdel (15) in 1990.

The Last Opinion of Laszlo Jamf

Pynchon's dark and malevolent scientist Laszlo Jamf is no longer, we read (1, p 577, line 7),

> In the last third of his life, there came over Laszlo Jamf—so it seemed to those who from out in the wood lecture halls watched his eyelids slowly granulate, spots and wrinkles grow across his image, disintegrating it toward old age—a hostility, a strangely *personal* hatred, for the covalent bond. A conviction that, for synthetics to have a future at all, the bond must be improved on—some students even read "transcended". That something so mutable, so *soft*, as a sharing of electrons by atoms of carbon should lie at the core of life, *his* life, struck Jamf as a cosmic humiliation. *Sharing?* How much stronger, how everlasting was the *ionic* bond—where electrons are not shared, but *captured. Seized!* and held! polarized plus and minus, these atoms, no ambiguities … how he came to love that clarity: how stable it was, such mineral stubbornness!

> "Whatever lip-service we may pay to Reason," he told Pökler's class back at the T.H., "to moderation and compromise, nevertheless there remains the lion. A lion in each one of you."

And so it would appear … he lives on.

References

1. Pynchon, Thomas. *Gravity's Rainbow;* Viking Penguin: New York, 1973.
2. Jacket blurb (unattributed) comparing Pynchon and Joyce.
3. Fowler, Douglas. *A Reader's Guide to* Gravity's Rainbow; Ardis Press: Ann Arbor MI, 1980.
4. Traced by Dietrich-Buchecker, C. O.; and Sauvage, J.-P. *Chem. Rev.* **1987**, *87*, 795.

5. Wasserman, E. *J. Am. Chem. Soc.* **1960**, *82*, 4433.
6. Frisch, H. L.; Wasserman, E. *J. Am. Chem. Soc.* **1961**, *83*, 3789.
7. Schill, G. *Catenanes, Rotaxanes, and Knots*; Academic Press: New York, 1971.
8. Walba, D. M.; Richards, R. M.; Haltiwanger, R. C. *J. Am. Chem. Soc.* **1982**, *104*, 3219.
9. Schill, G.; Lüttringhaus, A. *Angew. Chem.* **1964**, *76*, 567; and Schill, G. *Chem. Ber.* **1967**, *100*, 2021.
10. References 47 and 49, in Dietrich-Buchecker, C. O.; Sauvage, J.-P. *Chem. Rev.* **1987**, *87*, 795.
11. Dietrich-Buchecker, C. O.; Hemmert, C.; Khémiss, A.-K.; Sauvage, J.-P. *J. Am. Chem. Soc.* **1990**, *112*, 8002.
12. Frisch, H.; Martin, I.; Mark, H. *Monatsh. Chem.* **1953**, *84*, 250.
13. Patat, F.; Derst, P. *Angew. Chem.* **1959**, *71*, 105.
14. McConnell, H. *J. Chem. Phys.* **1961**, *35*, 508.
15. Farazdel, A.; Dupuis, M.; Clementi, E.; Aviram, A. *J. Am. Chem. Soc.* **1990**, *112*, 4205.

Bibliography

Helpful critical works on Pynchon and especially *Gravity's Rainbow* include

Mindful Pleasures: Essays on Thomas Pynchon; Levine, George; Leverenz, David, Eds.; Little, Brown: Boston, MA, 1976.
Critical Essays on Thomas Pynchon; Pearce, Richard, Ed.; Hall: Boston, MA, 1981
Tanner, Tony. *Thomas Pynchon*; Methuen: London, 1982.
Approaches to Gravity's Rainbow; Clerc, Charles, Ed.; Ohio State University Press: Columbus, OH, 1983.
Neuman, Robert D. *Understanding Thomas Pynchon*; University of South Carolina Press: Columbia, SC, 1986.
Weisenburger, Steven. A Gravity's Rainbow *Companion*; University of Georgia Press: Athens, GA, 1988.
The journal *Pynchon Notes*, published by J. M. Kraft and K. Tololyan, Middletown, CT.

Sherlock Holmes

The Eccentric Chemist

James F. O'Brien

Editor's Preface

For those who may feel that including a chapter on the exploits of a fictional detective in a book nominally dedicated to *chemistry* and *science fiction* requires some justification, the editor offers the following pertinent comments:

The hard science fiction aficionado demands his stories deal with known science, logically extrapolated, applied to a problem or to describe a situation. Using that yardstick, we have our analytically minded Holmes, clearly well-grounded in the chemical sciences, applying that knowledge to an assemblage of data and extrapolating from it a logical answer to a problem. Certainly such a viewpoint makes the Sherlock Holmes stories at least as valid SF as the works of Tolkien, any of the Oz books, the tales of Conan the Barbarian, and those stories dealing with time- or faster-than-light travel, all of which, regardless of popular viewpoint, ought to be considered "fantasy". It might be more appropriate to make a comparison with Crichton's *Andromeda Strain,* where the core of this SF story is similarly dedicated to solving a problem by the analytical treatment of a mass of largely scientific data. Obviously, a consequence of this inclusive definition is to subsume a modest segment of the "mystery" field as a marginal sub-genre of the science fiction bailiwick.

Perhaps for the above reasons the science fiction field has always exercised a certain fascination for a number of respected detective/mystery story writers. The list might begin with Arthur Conan Doyle, who, in addition to his better known Sherlock Holmes mysteries wrote the Professor Challenger stories (including *The Lost World*), continue with John D. MacDonald (science fiction such as *Wine of the Dreamers* and *Ballroom of the Skies)* as well as a number of others, often writing in either or both fields under noms de plume. The highly respected William White, for many years a senior book editor for the *New York Times*, wrote mysteries under the pseudonym of H. H. Holmes and science fiction as Anthony Boucher. Certainly Isaac Asimov's and August Derleth's future detectives, Wendell Urth and Solar Pons, respectively, are in the mold of Sherlock Holmes.

The above points have been made elsewhere but rarely as succinctly as by the author Kingsley Amis in the book *Science Fiction: The Future* (Dick Allen, Editor, Harcourt Brace Jovanovich, Inc., NY, 1971, p 257) as follows:

> I should like to assert flatly that detective fiction and science fiction are akin. There is a closely similar exaltation of idea or plot over characterization, and some modern science fiction, like most detective fiction, but unlike the thriller, invites the reader to solve a puzzle. It is no coincidence—how could it be?—that from Poe through Conan Doyle to Frederick Brown the writer of the one will often have some sort of concern with the other.

The perimeters of the SF genre tend to be elusive and it is convenient (as was proposed in Chapter 1) to adopt very flexible ones. Certainly there is room under the SF umbrella for the widely admired and often cherished team of Holmes and Watson.

Perhaps most importantly, the editor takes full responsibility for including this material in the original symposium, adds that nary a soul questioned its inclusion, points out that the talk drew one of the larger audiences, urges the reader to accept the coverage as at least marginal SF, and invites the reader to find in it rewarding reading.

The first Sherlock Holmes story (1), *A Study in Scarlet*, was published in 1887. Today, more than 100 years later and with copyrights expiring (2), the character is one of the most recognizable in all of literature. Put a deerstalker on the head and a pipe in the mouth and the public automatically thinks, "Sherlock Holmes". Old movies run on television again and again. New movies are made with regularity. Plays are done all around the country, even the world. Respectable presses publish Sherlock Holmes journals (3). There are even two works entitled "Encyclopedia Sherlockiana" (4, 5).

Although limited to the 60 original stories written by Sir Arthur Conan Doyle, Sherlock Holmes buffs eagerly seek out new Holmes stories by would-be Doyles (6, 7). These stories are called "pastiches", and science fiction buffs are easy marks for even marginal literature. Aspiring authors frequently base their stories on one of the more than 100 cases mentioned by Doyle but not reported in full (8, 9). Of course, "stories about the stories" are also coveted. To this day Doyle's estate gets letters from people who wish to use Holmes's services (10, 11). Numerous Holmes societies exist in the United States and around the world. The pinnacle of achievement for a Sherlock Holmes buff is an invitation to be a "Baker Street Irregular", a group apparently as odd as Holmes's ragamuffin street urchins from whom it takes its name.

Doyle imbued Sherlock Holmes with several features that give the character this appeal. Most important of these is his amazing deductive ability. Holmes is more clever than we are, but not so clever as to be unreal. We can follow his reasoning, once it is explained to us, and we can see ourselves mastering his cerebral approach. Holmes comes by this ability honestly. Doyle himself was more clever than most of us. In 1912, upon walking into a ward in a hospital in Edinburgh, Doyle's first words to a woman there with a listless child were, "Madam, you must stop painting your child's crib." The chemist can make the connection between heavy metals in paint and the symptoms they produce upon ingestion. Doyle, like Sherlock Holmes, could think on his feet. In 1893 Doyle wrote that he was "constantly receiving letters from persons in distress". These letters dealt with such things as unsolved personal problems and disappearances and came from all around the world (11). Occasionally, he was able to be of assistance (12).

Doyle's most famous forensic effort was the George Edalji case. Edalji, a Parsee considered a black man, was the victim of prejudice. He was convicted in 1903 of killing his neighbors' horses, cows, and sheep. After 3 years in prison, the clearly innocent Edalji sent information pertaining to his case to the recently knighted Sir Arthur Conan Doyle. Moved by the obvious injustice of the conviction, Doyle devoted the first 8 months of 1907 to clearing Edalji. Doyle did no work on his own projects and covered all of his own expenses. His efforts, which included identifying the actual criminal, resulted in a pardon for Edalji (13).

In similar fashion Doyle applied himself to right the injustice done to Oscar Slater, who was wrongfully imprisoned for murder for 18 years. He wrote numerous letters to newspapers; to members of Parliament; and to Sir John Gilmour, the Secretary of State for Scotland. He attended the appeal hearing in 1927, and he paid Slater's defense costs out of his own pocket (14).

In 1892 Doyle dedicated the first collection of short stories, *The Adventures of Sherlock Holmes*, to Dr. Joseph Bell as his inspiration for Holmes (15, 16). Clearly a case can be made that Doyle himself was Holmes. Certainly his wife Jean thought he was (17).

Sherlock Holmes and Logic

In addition to his clever competence, Holmes's love of logic is particularly appealing to scientists. His commitment to logic as a way of life is seen in *The Sign of the Four*: "I never guess. It is a shocking habit—destructive to the logical faculty" (18), and "I cannot agree with those who rank modesty among the virtues. To a logician all things should be seen as they are, and to underestimate oneself is as much a departure from the truth as to exaggerate one's own powers" (19).

In fact Sherlock Holmes's love of logic may have interfered with the kind of love that most people consider transcendent in human emotions (20): "Love is an emotional thing, and whatever is emotional is opposed to that true cold reason which I place above all things. I should never marry myself, lest I bias my judgment."

Doyle's stories, known as the canon by buffs, are filled with examples of Holmes exercising his deductive powers, not just talking about them. For example, the very first words that he speaks to his new acquaintance, Dr. John H. Watson, are, "How are you. You have been in Afghanistan, I perceive." Watson's response sets the tone for the rest of the journey through the 60 stories: "How on earth did you know that?"

The impressed Watson tells us his opinion (21): "He was the most perfect reasoning and observing machine that the world has seen." Some readers might dispute this statement and make a claim for Holmes's older and smarter brother Mycroft. We do not see much of Mycroft; he is mentioned in only 4 of the 60 tales. He does seem to get the better of Holmes in the following exchange from *The Sign of the Four* (22):

Mycroft: Look at these two coming toward us.

Sherlock: The billiard maker and the other?

Mycroft: Precisely. What do you make of the other?

Sherlock: An old soldier, I perceive.

Mycroft: And very recently discharged.

Sherlock: Served in India, I see.

Mycroft: And a noncommissioned officer.

Sherlock: Royal Artillery I fancy.

Mycroft: And a widower.

Sherlock: But with child.

Mycroft: Children, my dear boy, children.

Watson: Come this is too much!

As a final example of Holmes at his deductive best, we have, from "Silver Blaze" (23), what has become the most famous incident in all of the canon.

Inspector Gregory: Is there any other point to which you would wish to draw my attention?

Holmes: To the curious incident of the dog in the night-time.

Gregory: The dog did nothing in the night-time.

Holmes: That was the curious incident.

Sherlock Holmes and the Scientific Method

The appeal of Sherlock Holmes to chemists and other scientists lies somewhat in his use of the scientific method. A careful reading of the canon reveals that Holmes knew and understood the concepts of theory and hypothesis. Early in his career he notes (24), "It is a capital mistake to theorize before one has data. Insensibly one begins to twist facts to suit theories, instead of theories to suit facts."

Much later he reveals that his commitment to theory is still intact (25):

Holmes: What do you think of my theory?

Watson: It is all surmise.

Holmes: But at least it covers all the facts. When some new facts come to our knowledge which cannot be covered by it, it will be time enough to reconsider it.

On 64 separate occasions Holmes refers to theory; he refers 29 times to hypothesis. In a number of cases Holmes mentions his "working hypothesis" (26). Here is a detective that a scientist can love.

Observation and deduction, then, are his main tools (27). Mix in study, preparation, and the legwork so distasteful to the indolent Mycroft, and you get "the world's first consulting detective" (28). By the time we meet Holmes, we learn that he has been hard at work preparing himself for his unique profession. He has, in fact, authored several monographs that deal with the logical detection of crime and the pursuit of criminals. One of these monographs (29) is used in his investigation of the blackmailing McCarthy in "The Boscombe Valley Mystery". Undoubtedly, the monograph "The Typewriter and Its Relation to Crime" enabled Holmes to expose the cad Windibank in "A Case of Identity" (30).

A third monograph, "The Tracing of Footsteps, with Some Remarks upon the Use of Plaster of Paris as a Preserver of Impresses" (31) is of interest, because footprints are mentioned in 26 of the 60 stories. Holmes's place in the history of this forensic device has been recorded (32). His research for the monograph "The Influence of Trade upon the Form of the Hand, with Lithotypes of the Hands of Slaters, Sailors, Cork Cutters, Compositors, Weavers, and Diamond Polishers" (31) was of considerable use to Holmes, although he nearly errs in deducing the occupation of Violet Smith (33). In "The Red-Headed League", Holmes mentions his monograph "A Study of Tattoo Marks" (34), and he uses this knowledge to impress his client Jabez Wilson.

Perhaps his most famous monograph is "On the Use of Dogs in the Work of the Detective" (35). This monograph results in that wonderful scene in *The Sign of the Four* (36) where Holmes and Watson are led astray by Toby, the "ugly, long haired, lop eared creature" who, as Watson recounts, finds a creosote plant instead of a criminal:

"Here you are doggy! Good old Toby! Smell it Toby, smell it!"

He pushed the creosote handkerchief under the dog's nose. ... The creature instantly broke into a succession of high tremulous yelps and, with his nose on the ground ... pattered off upon the trail.

Figure 1. Toby leads them astray. (Reproduced with permission from reference 1. Copyright 1967 Clarkson N. Potter.)

I confess I had my doubts when I reflected upon the great traffic which had passed along the London road in the interval. My fears were soon appeased, however. Toby never hesitated or swerved. ... We had traversed Streatham, Brixton, Camberwell, and now found ourselves in Kennington Lane [Figure 1]. ... Toby ceased to advance but began to run backwards and forwards. ... Then he waddled round in circles.

"What the deuce is the matter with the dog?"

"Perhaps they stood here for some time", I suggested.

"It's all right, he's off again", replied Holmes.

Just past White Eagle Tavern ... the dog, frantic with excitement, turned ... down an alley, raced through sawdust and shavings, between two wood-piles and, with a triumphant yelp, sprang upon a large barrel. The staves of the barrel ... were smeared with a dark liquid and the whole air was heavy with the smell of creosote. Sherlock Holmes and I ... burst simultaneously into laughter.

Holmes, though, never lost faith in the ability of dogs (37): "Dogs never make mistakes."

Sherlock Holmes and the Formal Sciences

Holmes had considerable knowledge of the formal sciences. In the first story, *A Study in Scarlet*, Watson gives us this assessment of some of Sherlock Holmes's abilities (38):

- knowledge of literature, nil
- knowledge of philosophy, nil
- knowledge of astronomy, nil
- knowledge of politics, feeble
- knowledge of botany, variable
- knowledge of geology, practical
- knowledge of chemistry, profound

In "The Five Orange Pips" (39) and again in *The Sign of the Four* (40), Holmes makes use of his practical knowledge of geology. He is able to tell different soils at a glance and to associate them with different areas of London. This ability is very useful in tracking the movements of suspects.

We hear of Holmes's knowledge of physics in "The Adventure of the Six Napoleons"(41) when he says to Watson, "You will remember Watson how the dreadful business of the Abernetty family was first brought to my notice by the depth which the parsley had sunk into the butter on a hot day." Meteorology comes into play in "The Boscombe Valley Mystery" (42) when Holmes says, "It is entirely a matter of barometric pressure." Lestrade looked startled, "I do not quite understand."

Holmes had a working knowledge of both anatomy and botany. Watson appraises his knowledge of anatomy as "accurate, but unsystematic" (38). As usual, this knowledge took a practical bent. For example, Holmes first mentions fingerprints in *The Sign of the Four* and again in "The Man with the Twisted Lip". Both of these stories were published in the 1890s. Scotland Yard, however, did not adopt fingerprinting for identification until 1901.

As for botany, Watson touts Holmes as "well up in poisons, knows nothing of practical gardening" (38). Poisons occur in 22 of the 60 stories. Indeed, Doyle's first published medical article dealt with poisons (43).

Astronomy was an entirely different matter. In *A Study in Scarlet* (44) Watson says, "My surprise reached a climax when I found that he was ignorant of the Copernican Theory." Some evidence suggests that Watson underestimated Holmes's astronomy ability. For example, at one point Holmes discourses on "the obliquity of the ecliptic" (45).

Sherlock Holmes and Chemistry

Holmes mastered those parts of science that he felt would aid his forensic efforts. His knowledge of chemistry was to prove the most useful to him of all of the sciences. A "first-class chemist" is how young Stamford describes Holmes as he prepares to introduce Watson (46). Watson's own assessment of Holmes's chemistry initially was "profound" (38). After 6 years of living together, however, he downgraded the rating to "eccentric" (47). Isaac Asimov went so far as to refer to Holmes as "the blundering chemist" (48).

Abundant evidence confirms that Holmes was an avid chemist. A few quotes from the stories make this very clear. In "The 'Gloria Scott'" (49), which Holmes solved while a college student: "All of this happened during the first month of the long vacation between semesters. I went up to my London rooms, where I spent seven weeks working out a few experiments in Organic Chemistry." In *The Sign of Four* (50), Holmes unwinds, "I gave my mind a thorough rest by plunging into a chemical analysis." Interestingly, Holmes was much like today's students who often, I have heard, spend their spring-break vacations relaxing by doing chemistry, many of them at Padre Island, Texas.

Watson's view of Holmes's laboratory technique is not always positive (51), "Holmes ... busied himself all evening in an abstruse chemical analysis which involved much heating of retorts and distilling of vapors, ending at last in a smell which fairly drove me out of the apartment."

Even though logic may have been Holmes's first love, chemistry clearly was second, as was evident to Watson (52):

I found Sherlock Holmes half asleep, with his long form curled up on his armchair. A formidable array of bottles and test tubes with the smell of hydrochloric acid told me that he had spent the day in the chemical work that was so dear to him.

In "The Adventure of the Dancing Men", a story in which Holmes solves a cipher in order to solve the crime, he is even willing to interrupt a criminal investigation for a chemical investigation (53):

> If there is an afternoon train to town, Watson, I think we should do well to take it, as I have a chemical analysis of some interest to finish, and this investigation draws rapidly to a close.

Litmus

In addition to these quotes about Sherlock's chemistry, we are privileged to see him in action. Much of his work involves analytical chemistry. Early on he applies the use of litmus paper to the pursuit of a criminal (54):

> Holmes was seated at his side table ... working hard over a chemical investigation. In his right hand he had a slip of litmus paper.

> "You come at a crisis, Watson. If this paper remains blue, all is well. If it turns red, it means a man's life."

The outcome meant the man's life.

Acetones

One of Asimov's criticisms (48) of Holmes as chemist deals with a comment made in "The Adventure of the Copper Beeches" (55). In this story Watson reports (56):

> Holmes was settling down to one of his all night chemical researches which he frequently indulged in, when I would leave him stooping over the retort and a test tube at night, and find him in the same position when I came down to breakfast in the morning.

Subsequently in the same story (57) Holmes says, "Perhaps I'd better postpone my analysis of the acetones."

Asimov, pointing out that only one acetone exists, claimed that Holmes's use of the plural suggests an incompetence in chemistry. This criticism can be responded to in at least two ways: Holmes may have been dealing with a series of substituted acetones, or the common usage of the time was different than today. Evidence supports the possibility of a different common usage. An 1885 book on organic chemistry makes use of the exact same terminology as Holmes (58):

> By replacement of two hydrogen atoms of a paraffin on one and the same intermediate carbon atoms, there result derivatives ... whose oxygen compounds ... are termed ketones or acetones.

Asimov probably was not familiar with 19th-century chemical terminology. Using the plural of acetone was not necessarily a chemical blunder.

Coal Tar Derivatives

Apparently the only person in England who tired of Holmes was Doyle. He wrote (59) to his mother, "I think of slaying Holmes." She, an ardent Holmes fan, wrote back, "You won't, you can't, you mustn't."

Persuaded by his mother's plea (60), Doyle instead turned a plot suggestion she had made into "The Adventure of the Copper Beeches". Eventually though, Holmes had to go so that Doyle's time could be spent on what he considered more serious literary efforts. Having just visited the Reichenbach Falls in Switzerland, Doyle, in "The Final Problem", sends Holmes and Professor Moriarty over the precipice (Figure 2). In the resulting uproar 20,000 subscriptions to *The Strand* magazine were canceled (61). Doyle was inundated with letters and telegrams. His favorite, he always claimed (62), was from a lady who wired him simply, "You brute."

Not for 10 years would there be another Sherlock Holmes story; and when Doyle brought him back, circulation of *The Strand* soared by 30,000 (63). We learn in "The Adventure of the Empty House" that Holmes roamed the world for 3 years. Part of this time, which is known as "the great hiatus", was spent in the south of France doing research into coal tar derivatives (64).

Surmising why Sherlock Holmes would spend considerable time working on coal tar derivatives is not difficult. From antiquity people have sought to color their clothing. The most prized of the early natural dyes was Tyrian purple (65). The Phoenicians, in what is now Lebanon, extracted this material in minute amounts from a mollusk, *Murex brandaris*, found in the eastern Mediterranean. This expensive dye resulted in the city of Tyre acquiring a deserved reputation for odor due to the enormous quantities of rotting shellfish left on its shores. Now known to be dibromoindigotin (Figure 3), the extract was so expensive that only the very rich could afford to be garbed in purple. In fact, the term *porphyriogenatos*, literally "born in the purple", came to mean an emperor who was

Figure 2. Holmes and Moriarty plunge into the Reichenbach Falls. (Reproduced with permission from reference 110. Copyright 1987 Carroll and Graf.)

the son of an emperor (66). In the first century A.D. the Roman emperor Nero issued a decree that *only* the emperor could wear "royal" purple garments (67).

By Holmes's time little had changed. There were still only natural dyes: still very expensive. Importation of such dyes had cost England two million pounds in 1856. Then in London, William Henry Perkin, at age 18, oxidized what he thought was pure aniline and produced a purple product that he called aniline purple and eventually "mauve" (68). Perkin's discovery was actually an example of "serendipity". He had set out to synthesize quinine, which has the formula $C_{20}H_{24}N_2O_2$. Knowing no structural chemistry, he reasoned that if he could assemble a molecule with the correct

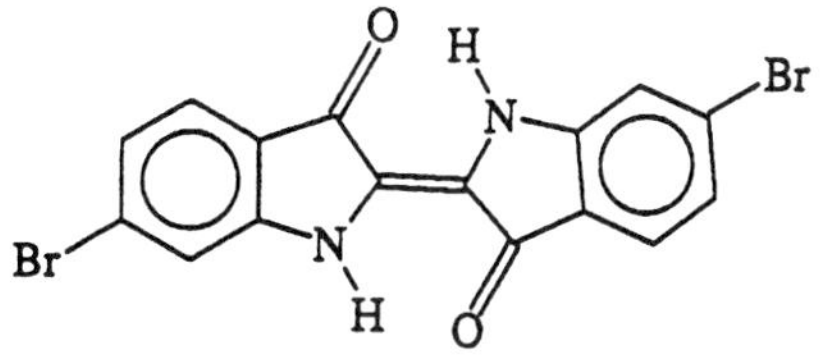

Figure 3. Dibromoindigo.

numbers of each atom, quinine would have to be the result (69). Accordingly Perkin attempted to remove a mole of water from the presumed oxidation product of N-allyltoluidine ($C_{10}H_{13}N$) using three moles of oxidizing agent. Little did he know that the structural formulas were so different. Instead of quinine, Perkin obtained an intensely colored red–brown precipitate. Fascinated by the unsuccessful reaction, Perkin repeated his experiment with a simpler base, aniline. His aniline, however, was contaminated by toluidine, and mauve ($C_{27}H_{25}N_4{}^+Cl^-$) was the result. He marketed mauve and related compounds as dyes. A large textile industry was already in place to use these new dyes. Perkin made so much money that he was able to retire in 1874 at age 36 and devote the remaining 33 years of his life to pure research (70).

The early years of English domination of the dye industry soon gave way to German pre-eminence. By 1881 50% of dyes were made in Germany. By 1900 fully 90% of dyes were made in Germany. This decline of English presence in dyes has been chronicled nicely (65). Some viewed the shift of the dye industry to Germany as lamentable. The noted British educator and chemist, Henry Enfield Roscoe, was quoted in 1881 (71):

> To Englishmen it is a somewhat mortifying reflection that whilst the raw
> materials from which all these coal tar colours are made are produced in
> our country, the finished and valuable colours are nearly all manufactured
> in Germany.

Meanwhile the German view of the dye situation was altogether different. The preface to an 1892 book by Theodore Weyl states (72):

> Thanks to the cooperation of theory and practice, the coal tar industry of
> Germany has conquered the world, and inasmuch as new and improved
> methods are continually being devised, will be able to maintain its pre-
> eminent position.

Recent scholarship (73) has linked a rising tide of German nationalism with chemical achievement, including German success in dyes as well as

Kekulé's disputed dreaming of the benzene structure. Therefore, a topic with "medico-legal" implications would naturally attract the attention of Sherlock Holmes. The patriotism of Holmes is well known. For example the very last story reported by Watson has Holmes coming out of retirement to expose the German spy von Bork on the eve of World War I. All the evidence points to the conclusion that, during the great hiatus, Holmes's research into coal tar derivatives dealt with the dye industry and England's efforts to stem the tide of German industry.

Holmes does not mention exactly what his efforts entailed, but scholars have made a strong case for his involvement with the synthetic dye industry (74). In his work with coal tar derivatives, Holmes probably dealt with the aforementioned creosote, which is a coal tar distillate. A major component of creosote is acenaphthene. Recently a suggestion was made that Holmes specifically worked with acenaphthene (75). As Inman pointed out, in *The Sign of the Four* Holmes is attempting to find a solvent in which to dissolve a hydrocarbon. Finding a solvent in which to dissolve a hydrocarbon is such a trivial task that Inman is able to make the case that what Sherlock actually sought was a solvent suitable for recrystallizing his hydrocarbon. This explanation may not be obvious, but as you can read in *The Merck Index* (76), acenaphthene can be recrystallized from ethanol.

Why did Holmes want to recrystallize acenaphthene? Again it was his interest in the dye industry. Inman pointed out the sequence of reactions that result in formation of a violet dye, a synthesis not published until 1908 (77). Inman made a strong case that Holmes found Grob's violet dye 15 years before its synthesis was published in *Berichte*.

Baryta

Sherlock Holmes also did some inorganic analytical work. At a lull in one case (78) he spends some time relaxing his mind by doing a chemical analysis. Watson returns from medical duties and enters their quarters:

Watson: Well, have you solved it?

Holmes: Yes, it was the bisulfate of baryta.

Watson: No. No the mystery!

Baryta is an old time name for barium oxide, BaO. Holmes is claiming to have analyzed or prepared barium bisulfate. This synthesis is difficult, for the compound is not well known even today. For example, barium bisulfate is not found in the *Handbook of Chemistry and Physics*. The

Swedish chemist Berzelius, by cooling a solution of barium sulfate in warm concentrated sulfuric acid, claimed to have isolated a 1:1 complex, $Ba(HSO_4)_2$ (79). In 1921 Kendall and Davidson (80), studying freezing points of barium sulfate in sulfuric acid, disputed Berzelius's results and found only the 3:1 complex, $BaSO_4 \cdot H_2SO_4$. A 1931 study (81) using solubility and conductivity data supported Berzelius's claim that only $BaSO_4 \cdot H_2SO_4$ exists in such solutions.

Blood Chemistry

Blood analyses, still important in criminalistics, were even more so in the late 1800s when modern instrumentation was not available. We learn in the first Holmes adventure of his interest in blood chemistry. The initial meeting of Holmes and Watson takes place in the laboratory at St. Bart's hospital in London, where we encounter the rarely excited Holmes (82):

> *Holmes:* I've found it. I've found it. … I have found a reagent which is precipitated by haemoglobin and by nothing else. … No doubt you see the significance of this discovery of mine?

> *Watson:* It is interesting chemically no doubt, but practically—

> *Holmes:* Why, man, it is the most practical medico-legal discovery for years. Don't you see that it gives us an infallible test for blood stains?

The Sherlock Holmes blood test was capable of detecting one part in a million (82). The guaiacum test then in use, first described by Christian Schönbein in 1859, was only good to one part in 100,000 (83). In addition to the guaiacum test a number of other tests for blood had been suggested before Holmes designed his test. In fact, 11 such tests were published between 1829 and 1871. Holmes made a very specific claim that his test was superior to the guaiacum-based test then in use. A number of authors have written about the unpublished Sherlock Holmes blood test (84–90). A recent article (90) made a convincing case for addition of sodium hydroxide to denature hemoglobin A, followed by precipitation by saturated ammonium sulfate. This type of chemistry was a remarkable achievement by Holmes in 1881.

Current crime lab consultants (91) spray luminol onto the suspected area to show the presence of blood stains. This method is extremely sensitive. If the luminol test shows blood to be present, a further test is then performed to distinguish human from animal blood. This is the precipitin test, which distinguishes human and animal blood based on the antibodies present.

The Blue Carbuncle

Another of Asimov's criticisms in his "Blundering Chemist" article (48) dealt with Holmes's use of the name blue carbuncle in "The Adventure of the Blue Carbuncle". Blue carbuncle, claimed Asimov, is a contradiction in terms. While different varieties of garnet may be different colors, only the red ones are termed carbuncles. In fact, the name carbuncle is used only for the dark red almandine garnet (92). Almandine garnet, one of six distinct chemical compositions of garnet, is an iron aluminum silicate, $Fe_3Al_2(SiO_4)_3$ (92).

In addition, Asimov pointed out, Holmes refers to the blue carbuncle as "this 40 grain weight of crystallized charcoal". Holmes is confusing carbuncles with diamonds according to Asimov. I propose that Holmes was correct in both cases: in the use of the term carbuncle, and in the reference to diamond.

In the 19th century, garnet-topped doublets were common in cheap jewelry (93). In these stones a thin sliver of garnet was fused to the top of (usually) glass or sometimes another stone (Figure 4). The color of the garnet was not visible. In this way any gem could be mimicked, *any color produced*. Garnets were frequently used in making doublets because of their high luster, excellent durability, and easy availability. They do not crack when fused. The blue carbuncle was a blue doublet, referred to as a carbuncle because the thin top was almandine garnet.

What about diamond? Holmes's blue carbuncle consisted of a garnet fused to a diamond, likely a blue diamond. Such doublets, those made from two precious stones, are called "true doublets". Diamonds are not often used in doublets; after all they are the most valuable of stones. Why mask

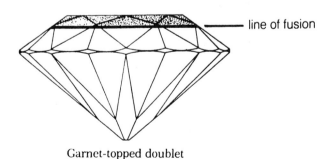

line of fusion

Garnet-topped doublet

Figure 4. A garnet-topped doublet. (Reproduced with permission from reference 108. Copyright 1989 Gemstone.)

their appearance! When diamonds are used in doublets, it is invariably with another diamond, the goal being to make it appear as one large diamond (94).

What could be a greater sign of affluence than to afford to cover up a diamond in a doublet stone. Absolutely no other similar stone existed. This combination made the "blue carbuncle" unique, and, somewhat paradoxically, gave it the great value placed upon it.

Eccentric, not Profound

Chemical topics are mentioned in 56 of the 60 tales. Despite this seemingly heavy involvement, Watson's assessment of Sherlock Holmes's chemistry as "eccentric" is more accurate than his original rating of "profound". Holmes was a chemical hobbyist rather than a real user of the science. Never does he solve a case by chemical means. In the first story, *A Study in Scarlet*, he raves to Watson, a new acquaintance, about the Sherlock Holmes blood test. He never uses it in a case. On several occasions Holmes stays up all night working on "chemical researches". Apparently these researches were not vital to his work. He was just the type of person who got extremely involved in what he was doing. For example, when his landlady Mrs. Hudson asks when he would like to eat, Holmes replies (95), "Seven thirty the day after tomorrow."

Holmes was a true eccentric, and not just in his chemical researches. Mrs. Hudson exhibited admirable tolerance for the behavior of the "very worst tenant in London" (96). Holmes kept cigars in a coal scuttle and tobacco in a Persian slipper (97), affixed his correspondence to the mantelpiece with a jack knife (97), exhibited "incredible untidyness" (96), honored his queen by shooting VR into the apartment wall with bullets (97), and performed the malodorous chemical experiments already mentioned. Watson even described Holmes's "powers upon the violin" as "eccentric" (98). It seems that Holmes "would seldom produce any music". Instead he preferred to play only chords, "sometimes melancholy", "occasionally fantastic" (97). All of this was done, of course, on his own Stradivarius (99).

Holmes occasionally was too much for Mrs. Hudson to deal with. At least one time a substitute landlady, Mrs. Turner, is called upon to give Mrs. Hudson a break from her famous tenant's odd behavior (100). In general, though, Mrs. Hudson coped very well with Holmes's enormous idiosyncrasies as well as the atmosphere of violence and danger that surrounded Holmes (96).

Of course, there is one more chemical that must be mentioned (*101*):

Dr. Watson: Which is it today, morphine or cocaine?

Sherlock Holmes: It is cocaine. A seven per cent solution. Would you care to try it?

Holmes was fortunate to have as a friend a physician such as Watson. The doctor, usually so deferential to Holmes, took a stand when it came to drug usage. He was willing to call Holmes a "self poisoner" (*102*). Never able to adjust to the sight of the syringe, whether it was on the mantel (*103*) or being inserted into Holmes's arm (*101*), Watson finally weaned Holmes from the habit (*104*).

Sherlock Holmes was truly eccentric. His knowledge of chemistry, his use of chemicals, and his entire lifestyle were eccentric. This very eccentricity has kept the character alive. He was, as Watson says in "The Final Problem", "the best and wisest man whom I have ever known" (*105*).

References

1. All references to the original Sherlock Holmes stories are to the two-volume set edited by W. S. Baring-Gould, *The Annotated Sherlock Holmes*; Clarkson N. Potter, Inc.: New York, 1967.
2. Peck, A. J. "Sherlock Holmes and the Law", *Baker Street Miscellanea* **1984**, *40*, 16–24.
3. For example, *The Baker Street Journal*, P.O. Box 465, Hanover, PA 17331; *The Sherlock Holmes Journal*, The Sherlock Holmes Society of London, Highfield Farm House, 23 Highfield Avenue, Headington , Oxford, OX3 7LR.
4. Tracy, J. *The Encyclopedia Sherlockiana*; Doubleday: New York, 1977.
5. Bunson, M. E. *Encyclopedia Sherlockiana*; Macmillan: New York, 1994.
6. Boyer, R. L. *The Giant Rat of Sumatra*; Warner Books: New York, 1976.
7. Doyle, A. C.; Carr, J. D. *The Exploits of Sherlock Holmes*; Random House: New York, 1952.
8. Baring-Gould, W. S. *Sherlock Holmes of Baker Street*; Bramhall House: New York, 1962; pp 296–317.
9. Reference 5, p xx, xxi.
10. Asimov, I. "Thoughts on Sherlock Holmes", *Baker Street Journal* **1987,** *37*, 201.
11. Costelleo, P. *The Real World of Sherlock Holmes*; Carroll and Graf Publishers: New York, 1991; p 25.
12. *The Sherlock Holmes Scrapbook*; Haining, P., Ed.; Clarkson N. Potter, Inc.: New York, 1974; p 25.
13. For a complete description of the Edalji case *see* Carr, J. D. *The Life of Sir Arthur Conan Doyle*; Vintage Press: New York, 1975; pp 268–290.
14. *Famous Trials*; Hodge, H.; Hodge, J. H., Eds.; Dorset Press: New York, 1986; pp 105, 131.

15. Bunson, M. E. *Encyclopedia Sherlockiana*; Macmillan: New York, 1994; p 4.
16. Green, R. L. Introduction, *The Adventures of Sherlock Holmes*; Oxford University Press: Oxford, England, 1993; pp xvii–xix.
17. Reference 12, p 29.
18. *The Sign of the Four* , W. S. Baring-Gould, Vol. 1, p 614.
19. Ibid., p 591.
20. Ibid., p 687.
21. "A Scandal in Bohemia", reference 1, Vol. 1, p 346.
22. "The Greek Interpreter", reference 1, Vol. 1, p 594.
23. "Silver Blaze", reference 1, Vol. 2, p 277.
24. "A Scandal in Bohemia", reference 1, Vol. 1, p 349.
25. "The Yellow Face", reference 1, Vol. 1, p 586.
26. Hudson, R. L. "Theory, Hypothesis and Sherlock Holmes", *Baker Street Journal* **1991,** *41,* 88.
27. Induction as well, *see* Byerly, A. "It Is the Scientific Use of the Imagination", *Baker Street Miscellanea* **1983,** *36,* 32–34.
28. *A Study in Scarlet*; reference 1, Vol. 1, p 160; *The Sign of Four*; Vol. 1, p 611.
29. *The Sign of Four*; reference 1, Vol. 1, p 612; *A Study in Scarlet*; Vol. 1, p 173; "The Boscombe Valley Mystery", Vol. 2, p 148;
30. "A Case of Identity", reference 1, Vol. 1, p 414.
31. *The Sign of Four*; reference 1, Vol. 1, p 612.
32. Vatza, E. J. "An Analysis of the Tracing of Footsteps from Sherlock Holmes to the Present", *Baker Street Journal* **1987,** *37,* 16.
33. "The Solitary Cyclist", reference 1, Vol. 2, p 384.
34. "The Red-Headed League", reference 1, Vol. 1, p 418.
35. "The Adventure of the Creeping Man", reference 1, Vol. 2, p 752.
36. *The Sign of Four*; reference 1, Vol. 1, pp 645–649.
37. "Shoscombe Old Place", reference 1, Vol. 2, p 638.
38. *A Study in Scarlet*; reference 1, Vol. 1, p 156.
39. "The Five Orange Pips", reference 1, Vol. 1, p 390.
40. *The Sign of the Four*; reference 1, Vol. 1, p 613.
41. "The Adventure of the Six Napoleons", reference 1, Vol. 2, p 574.
42. "The Boscombe Valley Mystery", reference 1, Vol. 2, p 140.
43. "Action of a Gelseminum Sempervireus", *Brit. Med. J.* **1879.**
44. *A Study in Scarlet*; reference 1, Vol. 1, p 154.
45. "The Greek Interpreter", reference 1, Vol. 1, p 590.
46. *A Study in Scarlet*; reference 1, Vol. 1, p 148.
47. "The Five Orange Pips", reference 1, Vol. 1, p 399.
48. Asimov, I. "The Blundering Chemist", *Sci. Dig.* **1980,** *88,* 9.
49. "The 'Gloria Scott'", reference 1, Vol. 1, p 112.
50. *The Sign of Four*; reference 1, Vol. 1, p 662.
51. Ibid., p 657.
52. "A Case of Identity", reference 1, Vol. 1, p 413.
53. "The Adventure of the Dancing Men", reference 1, Vol. 2, p 539.
54. "The Naval Treaty", reference 1, Vol. 2, p 169.
55. "The Adventure of the Copper Beeches", reference 1, Vol. 2, p 114.
56. Ibid., p 120.

57. Ibid., p 121.

58. *Adolph Strecker's Short Textbook of Organic Chemistry*, 2nd ed; J. Wislicenus: London, 1885; p 275.

59. Carr, J. D. *The Life of Sir Arthur Conan Doyle*; Vintage Press: New York, 1975; p 100.

60. Ibid., p 99.

61. Jaffe, J. A. *Arthur Conan Doyle*; Twayne Publishers: Boston, MA, 1987; p 9.

62. *The Standard Doyle Company: Christopher Morley on Sherlock Holmes*; Rothman, S., Ed.; Fordham University Press: New York, 1990; p 109.

63. Reference 1, Vol. 1, p 16.

64. "The Adventure of the Empty House", reference 1, Vol. 2, p 337.

65. Hudson, J. *The History of Chemistry*; Chapman and Hall: New York, 1992; p 4.

66. Salzberg, H. W. *From Caveman to Chemist: Circumstances and Achievements*; American Chemical Society: Washington, DC, 1991; p 3.

67. McGovern, P. E.; Michel, R. H. *Acc. Chem. Res*, **1990**, *23*, 152–158.

68. Saltzman, M. D.; Kessler, A. L. "The Rise and Decline of the British Dyestuffs Industry", *Bull. Hist. Chem.* **1991**, *9*, 7–15.

69. Lehman, J. W. *Operational Organic Chemistry*, 2nd ed.; Allyn and Bacon: Boston, MA, 1988; p 401.

70. Roberts, R. M. *Serendipity: Accidental Discoveries in Science*; John Wiley and Sons: New York, 1989; p 66.

71. As quoted in Cliffe, W. H. "A Historical Approach to the Dyestuffs Industry", *J. Soc. Dyers Colourists* **1963**, *79*, 353–363.

72. Weyl, T. *The Coal Tar Colours with Especial Reference to Their Injurious Qualities and the Restriction of Their Use, a Sanitary and Medico-Legal Investigation*; P. Blakiston: Philadelphia, PA, 1892.

73. Wotiz, J. H.; Rudofsky, S. "The Unknown Kekulé", in *Essays on the History of Organic Chemistry*; Traynham, J. G., Ed.; Louisiana State University Press: Baton Rouge, LA, 1987; p 30.

74. Caplan, R. M. "Why Coal-Tar Derivatives at Montpellier?" *Baker Street Journal* **1989**, *39*, 29.

75. Inman, C. G. "Sherlockian Distillates", *J. Chem. Educ.* **1987**, *64*, 1014–1015.

76. *The Merck Index*, 10th ed.; Merck and Co., Inc.; Rahway, NJ, 1983; p 29.

77. Grob, A. *Berichte.* **1908**, *41*, 3331–3334

78. "A Case of Identity", reference 1, Vol. 1, p 413.

79. Berzelius, J. J. *Ann.* **1843**, *46*, 250; Gillard, R. D. *Educ. Chem* **1976**, *13*, 10–11.

80. Kendall, J.; Davidson, A. W. *J. Am. Chem. Soc.* **1921**, *43*, 979.

81. Trenner, N. R.; Taylor, H. A. *J. Phys. Chem.*, **1931**, *35*, 1336.

82. *A Study in Scarlet*; reference 1, Vol. 1, p 150.

83. Sutherland, W. D. *Blood Stains: Their Detection and the Determination of Their Source. A Manual for the Medical and Legal Professions*; Baillere, Tindall, and Cox: London, 1907.

84. Pinkus, J. L.; Goldman, L. S. "A Benzidine Rearrangement and the Detection of Trace Quantities of Blood", *J. Chem. Educ.* **1977**, *54*, 380.

85. Gillard, R. D. "Sherlock Holmes—Chemist", *Educ. Chem*, **1976**, *13*, 10.

86. McGowan, R. J. "Sherlock Holmes and Forensic Chemistry", *Baker Street Journal* **1987**, *37*, 10–14.

87. Redmond, D. A. "Some Chemical Problems in the Canon", *Baker Street Journal* **1964**, *14*, 145–152.

88. Holland, V. R.; Saunders, B. C.; Rose, F. L.; Walpole, A. L. "A Safer Substitute for Benzidine in the Detection of Blood", *Tetrahedron* **1974,** *30*, 3299.
89. Adler, O.; Adler, R. *Zeit. Phys. Chem.* **1904,** *41*, 59.
90. Huber, C. L. "The Sherlock Holmes Blood Test: The Solution to a Century Old Mystery", *Baker Street Journal* **1987,** *37*, 215–220.
91. Reta Tindall, Forensic Chemist, Pinellas County Forensic Laboratory, Florida.
92. Sinkankas, J. *Gem Cutting,* 2nd ed.; Van Nostrand: New York, 1962.
93. Rutland, E. H. *An Introduction to the World's Gemstones;* Doubleday: Garden City, NY, 1974.
94. Matlins, A. L.; Bonanno, A. C. *Gem Identification Made Easy;* Gemstone Press: South Woodstock, VT, 1989.
95. "The Mazarin Stone", reference 1, Vol. 2, p 735.
96. "The Dying Detective", reference 1, Vol. 1, p 439.
97. "The Musgrave Ritual", reference 1, Vol. 1, p 123.
98. *A Study in Scarlet;* reference 1, Vol. 1, p 158.
99. "The Cardboard Box", reference 1, Vol. 2, p 200.
100. "A Scandal in Bohemia", reference 1, Vol. 1, p 361.
101. *The Sign of the Four;* reference 1, Vol. 1, p 610.
102. "The Five Orange Pips", reference 1, Vol. 1, p 399.
103. "The Dying Detective", reference 1, Vol. 1, p 442.
104. "The Missing Three-Quarter", reference 1, Vol. 2, p 475.
105. "The Final Problem", reference 1, Vol. 2, p 317.
106. Reference 1, Vol. 2, p 488.
107. Bunson, M. E. "Encyclopedia Sherlockiana", Macmillan: New York, 1994 ; p. 155.
108. Matlins, A. L.; Bonnano A. C. *Gem Identification Made Easy;* Gemstone Press: South Woodstock, VT, 1989; p 169.
109. Reference 1, p. 206.
119. *The New Adventures of Sherlock Holmes;* Greenberg, M. H.; Rossel Waugh, C. L., Eds.; Carroll and Graf Publishers: New York, 1987; p 132.

⋆ 8 ⋆

The Right Environment

Using Real Chemistry as a Basis for Science Fiction

Harry E. Pence

⋆The right environment can be an important component for any work of fiction, but it is especially important for science fiction. This discussion will focus on the ways in which two different authors used their conceptions of the environment to increase the interaction of the reader with the story. The term "environment" can have many different meanings, but the context here will focus on three different aspects of this general term.

First, the obvious way to think about the environment is the combination of known scientific facts and reasonable extrapolations that serve as a background for the development of the plot. This background can be very important, because inconsistencies or violation of scientific fact undermine the reader's involvement and so his or her enjoyment of the story. Creating an environment that is internally consistent, but is totally contrary to current scientific knowledge, is an approach that is the basis for the very popular field of science fantasy.

The second aspect of environment has been described by critics who write *about* science fiction, such as Mark Rose, who suggested that the environment plays a much different role. Rose argued (1) that the essential metaphor of science fiction is a mortal combat between two antagonists:

people's power to comprehend and nature's power to destroy. The first is essentially spiritual and the second, physical, so that conflict may not be specifically resolved in the story. Because the spiritual transcends the physical, the effort to understand and control nature is assumed to be more important than the survival of the individual humans involved.

A third way to look at the environment is to ask how well the plot corresponds to the attitudes and accepted practices associated with real science. Two of the major characteristics of science are problem-solving and cooperative efforts. Therefore, the right environment should reflect these observations from the real world; problem-solving and cooperation should play an important role in the plot.

The two writers whom I have chosen for this discussion deal with these three aspects of the environment in different ways, but both attempt to define the right environment.

The Moon Is Hell

Any discussion of using a realistic environment as a basis for science fiction must, almost inevitably, begin with John W. Campbell, Jr. As editor of *Astounding Stories* (later renamed *Astounding Science Fiction*, and finally *Analog*) from 1937 to 1971, Campbell played a major role in the development of science fiction (2). During this period, *Astounding* focused especially on fact-based stories, so much so that Brian Aldiss suggested (3) that, "There were times when *Astounding* smelt so much of the research lab that it should have been printed on filter paper."

Although he is remembered today primarily as an editor rather than as an author, Campbell wrote a number of novels and stories, the best known of which, "Who Goes There?", written in 1938 under the pseudonym of Don A. Stuart, became the basis for two well-known science fiction movies. His early writings are best described as "space operas"; plot and characterization were simple, and the emphasis was on superweapons and rapid action. *The Moon Is Hell*, published in 1951, represents an exception and was, for its time, an unusually realistic attempt to portray space travel.

This novella is essentially a modern-day version of the Robinson Crusoe story. The United States has created a station on the dark side of the Moon, and the crash of a space shuttle carrying replacement supplies leaves the 13 members of the expedition (all male) stranded with little food, water, or oxygen. The story, told mainly through the diary of one of the participants, tells of the efforts by the members of the expedition to use only their own resources to survive until another relief ship arrives.

The desolate lunar landscape becomes, as the title implies, a major character in the story; the scientists must overcome nature itself to fulfill their crucial needs. They find a large deposit of gypsum near the space station, mine it, and heat the mineral to obtain water. A bed of silver selenide provides silver, which can be used to make photocells, which are, in turn, a source of current to electrolyze the water to hydrogen and oxygen. Recombination of the hydrogen and oxygen produces a source of heat during the lunar night. Systematically, the stranded scientists satisfy their basic needs by winning them from the apparently sterile lunar surface.

The next difficulty encountered is the diminishing food supply, and one of the chemists in the group discovers how to synthesize replacement food supplements from the combination of basic elements. Water is obtained from gypsum, as previously described. Nitrogen and carbon dioxide are derived from minerals containing nitrides and carbides. Although the group is able to synthesize or find all of the known essential nutrients and vitamins, they are not aware that a new vitamin has been discovered on Earth. The lack of this essential component leaves most of the surviving members of the expedition near death by the time the relief ship arrives.

Campbell has obviously taken unusual care to make his story correspond to the science of his day, and much of it holds together quite well. Using gypsum as a source of water, making the photoelectric cells, even synthesizing food replacements from inorganic starting materials seems at least possible, if not always achievable. Campbell's biggest failure was, as he would have probably guessed himself, his inaccurate picture of the lunar surface.

During the 1960s and 1970s, astronauts from the United States and the then Soviet Union visited a number of lunar sites and brought back rock samples that provide an interesting contrast with the situation predicted in Campbell's story. Lunar rocks are currently thought to have crystallized under very dry, extremely reducing conditions, and the volatile elements were apparently depleted prior to aggregation. The lack of water would prevent some of the most important processes responsible for unusual concentrations of useful elements in ores, and frequent impacts by meteorites on the lunar surface have probably caused the composition to be unusually uniform. In short, much of what Campbell hypothesized about the Moon was wrong.

Thus, many of the materials that Campbell suggested would be discovered and used by his fictional lunar colony are apparently not available.

Compared to terrestrial abundances, Mason and Melson (4) reported low levels of silver, and therefore it is unlikely that silver could be used to create photocells. Most important, Mason and Melson indicated that there is little possibility of finding the gypsum deposits that Campbell suggested were so essential as a source of water, and also the concentrations of nitrogen and carbon are quite low. A number of lunar rocks do contain oxygen, but simple heating in a vacuum will not liberate much of this essential gas.

Despite these inaccuracies, Campbell was very successful at creating an environment for the action. For the time when it was written, the science is quite logically consistent and reasonable. The focus rarely shifts from the essential conflict. A subplot concerns one of the expedition who first steals food, then attempts to kill the remaining scientists, but this subplot is treated in an almost perfunctory way. Although not dealing specifically with this novel, Rose (5) described the situation well when he wrote, "What is important is the attempt to replenish the void, to fill the immense absence with meaning, even if this is accomplished by turning emptiness itself into an antagonist that can be confronted in human terms."

Campbell's portrayal of the attitudes that characterize science has held up over time much better than his scientific extrapolations. He emphasized the two characteristic attributes, problem-solving and cooperation. Indeed, the main plot consists almost totally of the attempts by the members of the lunar expedition to solve a succession of problems in order to obtain the basic necessities of life from their cruel and unforgiving surroundings.

Cooperative efforts also play an important role. Even though the chemist in Campbell's lunar novel is a rugged individualist, he can succeed only if he obtains help from his friends and co-workers. The lesson is clear: There are times to work alone, but there is also a real need to be willing and capable of joining with a group and making a joint contribution. The man who steals food violates this need to cooperate, and this failure eventually leads to his death.

Mining the Moon for Oxygen

There may be practical ways to mine the moon for oxygen. In a recent article, Burt (6) discussed several ways of obtaining oxygen on the moon and concluded that it might not only be possible but even profitable. Burt pointed out that its low gravity, proximity, and suitability as a base for further exploration make the moon a good prospect for mining. He proposed mining for the elements most common on the moon: silicon, aluminum, calcium, iron, titanium, magnesium, and oxygen. Of these, he suggested that oxygen would probably be the most valuable.

On the basis of Burt's description, some aspects of Campbell's story do seem realistic. The low gravity would make ore transport easy. Surface mining is probably feasible, because the lunar soils are already pulverized and would not require crushing or grinding. Electrostatic concentration methods, such as those used by the stranded scientists, should work better than more traditional benefaction methods, such as flotation or gravity concentration. Unfortunately, there is one major difference; Burt indicated that the smelting and refining process could be performed only by using reagents brought to the moon for that purpose.

On closer examination, Burt's proposals for obtaining oxygen do not seem completely practical, even if the reagents are brought from Earth. For example, one suggestion is to use hydrogen to partially recover the oxygen from ilmenite, $FeTiO_3$, which is present in the basalt rocks found on the maria.

$$FeTiO_3 + H_2 \rightarrow Fe + TiO_2 + H_2O$$

In addition to the requirement that the hydrogen must be provided from the Earth, this process has other disadvantages, including the fact that it is barely possible on the basis of thermodynamics; only one-third of the available oxygen is recovered, and the oxygen is recovered in the form of water, which must then be electrolyzed.

Burt also discussed a more practical method for obtaining oxygen, involving the combination of fluorine gas with ilmenite or anorthite ($CaAl_2Si_2O_8$), followed by reaction with sodium. This method would have the advantages that the oxygen is directly produced as a gas, and several useful by-products would result, including aluminum, silicon (for photocells), and calcium oxide (for use in ceramics, cements, or batteries). The disadvantage is that the fluorine would need to be brought from Earth, probably in the form of sodium fluoride, and producing the fluorine by electrolysis would be very energy-intensive.

T. P. Caravan and the Evil Chemistry Professor

During the early 1950s, a set of short stories that presented a different view of science appeared in *Other Worlds Science Stories*. Unlike *Astounding*, *Other Worlds* has largely been forgotten. It was one of several science fantasy magazines created by Ray Palmer, whose publications specialized more in space opera than in sophisticated fiction. *Other Worlds* was published from 1949 to 1957, and was eventually transformed into a magazine about flying saucers (7).

The author of these stories, T. P. Caravan, was unusual in that he attempted to combine humor with science. A search of standard sources revealed little information about Caravan, except for a cryptic note that precedes one of his stories in an anthology published in 1963 (8):

> Mr. Caravan is an extremely shy man. He is on the faculty of the College of the City of New York, and is an authority on 18th Century English literature. Once he was an aerial gunner, but now he's too fat.

It seems quite possible that the name is a pseudonym.

Four stories in this series have been located, published in the period 1952 to 1953. The main protagonist in each of these stories is John, a young science student, who is described as being a genius, but no one ever knows it because he never does his homework. He is continually having problems with the old and evil professor of chemistry. To complicate matters, John loves the professor's beautiful daughter. This situation makes the professor, who loves to flunk students at any time, even more eager to eliminate poor John. The series of short stories describes how John manages to survive various predicaments created by the professor, weds the professor's beautiful daughter, and presumably graduates (although that is not described in any of these stories).

In "Happy Solution" (9), the professor agrees to give his permission for John to marry his daughter, if John can fulfill three impossible conditions. He must bring the professor a lock of hair from the dean of the college (who is totally bald), he must get an A in chemical thermodynamics (a feat that no one had ever accomplished in the "ninety-eight point six" years that the evil professor has been teaching the course), and, finally, he must provide the professor with a container that would hold his universal solvent. John borrows the lock of hair that the dean's wife has kept as a souvenir, obtains the impossible A in thermodynamics (with help from a flyspeck on the grade report and a relative in the registrar's office), and solves the problem of the universal solvent by freezing it in a block of ice.

The other stories are in a similar vein. "Last Minute" (10) describes how John creates a perpetual motion machine (when he should have been doing his thermodynamics homework) and absentmindedly decorates it with some diamonds that were sitting on the professor's desk. The professor promises to have John expelled unless he can recover the diamonds, which is, of course, impossible, because the machine will run forever. John cools the machine to absolute zero and causes time to stop long enough to retrieve the stones.

In "Dinosaur Day" (*11*), the campus is threatened by a tyrannosaurus that the evil professor has brought from the past by means of a time machine. This time the solution is provided by John's wife (formerly the professor's daughter), who reasons that because dinosaurs died out as a result of the Ice Age, all she has to do is turn up the air conditioning, and the beast will die of the cold. It works, and the professor is foiled again.

The last of the four stories is "The Cold, Cold Grave" (*12*). The mad professor tricks John into taking a solution that will kill "the smallest cells in the human body". The professor assumes that killing all of any cell type must result in death, but he does not notice that John has a cold. The cold viruses are the smallest cells in John's body, so instead of killing John, the professor has found a cure for the common cold. The professor is enraged, John is relieved, and the people of the world are delighted that they no longer will suffer from colds.

Caravan's stories may seem like unlikely choices for an article concerned with the need to establish the right environment, but close examination reveals some interesting similarities with Campbell's lunar narrative. Caravan uses scientific ideas as the basis for his stories; however, the ideas are observed mainly by their violation. Perpetual motion machines, universal solvents, time travel, and grades of A in chemical thermodynamics are all impossible (or at least unlikely) according to our current knowledge, but of course the humor is made possible by the author and the reader sharing the knowledge that these are improbable.

Both authors also deal with the importance of problem-solving and cooperation in science. Caravan makes problem-solving the main theme of his stories, but most of the solutions violate natural law. John must solve problems to marry the evil professor's daughter, or learn how to stop a perpetual motion machine, or save the campus from the dinosaur. Only "The Cold, Cold Grave" breaks this pattern, and here the problem-solver is the professor. He does not think carefully through the question of what might be the smallest cells in the human body, and therefore his attempt at problem-solving fails, much to John's relief.

Cooperation is also an underlying theme in these stories. When John is ready to accept that he cannot find any way to kill the dinosaur, his wife solves the problem. The need for cooperation is also emphasized in a negative way. The scientific ability of the old and evil professor of chemistry is never questioned, but his isolation from society makes him ridiculous.

It is more difficult to argue that these stories demonstrate Mark Rose's idea that the environment is a major protagonist in science fiction. One might suggest that nature participates to the extent that its fundamental laws are being violated with alarming consistency, but even that may be stretching things a bit too far. Despite this argument, it is easy to see that there are many resemblances between T. P. Caravan's approach to science fiction and Campbell's approach. The main difference is that Caravan shows that it is possible to make fun of even the most staid and respectable concepts, an idea that may be lost too often in the seriousness of modern science.

The Historical Environment

I have argued that an important factor in the success of a science fiction story is the ability of the author to create a realistic environment as a setting for the plot and then to have that environment make an effective contribution to the development of the story. A related problem is, How does the historical environment at the time affect the author as he or she is writing the story? To put it another way, have the societal changes since the 1950s affected the right environment?

Science fiction is as much about the present as it is about the future; often the author is commenting on some particular current development by extrapolating it into the future and showing that it ultimately leads to disaster or else demonstrating that flaws in the present society will make it unable to cope with a significant challenge, for example, alien invasion or climate change. If this is true, then the plot environments that were appropriate for the 1950s should be quite different from those appropriate today.

The stories discussed here seem to reflect a pervasive optimism about the future and a firm belief that technology can solve all problems. As Helen Parker pointed out (13), this sort of hopefulness is relatively uncommon among modern science fiction writers. Some writers even view problem-solving as an inappropriate focus for their stories. For example, Barry Malzberg, one of the more successful recent science fiction writers, recounted (14) that as a young man he had the following conversation with John Campbell:

> "You've got to understand the human element here," the young man said, "it's not machinery, it's people, people being consumed at the heart of these machines, onrushing technology, the loss of individuality, the

loss of control, *these* are the issues that are going to matter in science fiction for the next fifty years. It's got to explore the question of victimization."

"I'm not interested in victims," Campbell said, "I'm interested in heroes. I have to be; science fiction is a problem-solving medium, man is a curious animal who wants to know how things work and given enough time can find out."

"But not everyone is a hero. Not everyone can solve problems." …

"Those people aren't the stuff of science fiction," Campbell said. "If science fiction doesn't deal with success or the road to success, then it isn't science fiction at all."

Thus, choosing the right environment seems to be much more complex than might be apparent. Perhaps the only real test is whether or not the environment is sufficiently realistic to convince the reader to suspend disbelief and participate in the narrative. The stories discussed here may not qualify as great works of literature, but they were convincing enough so that at least one reader clearly remembered them 40 years later. Judging from that basis, it would appear that they did, indeed, have the right environment.

Acknowledgments

I thank the reference staff at Milne Library, State University of New York at Oneonta for their enthusiastic assistance. I also express my appreciation to Jack Stocker, who worked so hard to make the symposium upon which this book is based a reality.

References

1. Rose, Mark. *Alien Encounters;* Harvard University Press: Cambridge, MA; 1981.
2. *Science Fiction Writers;* Bleiler, E. F., Ed.; Charles Scribner's Sons: New York, 1982.
3. Aldiss, Brian W. *Trillion Year Spree;* Avon Books: New York; 1986, p 225.
4. Mason, Brian; Melson, William G. *The Lunar Rocks;* Wiley-Interscience: New York; 1970.
5. Reference 1, p 53.
6. Burt, Donald M. "Mining the Moon", *Am. Sci.* **1989,** *77,* 574-578.

7. *The Science Fiction Encyclopedia*; Nicholls, Peter; Clute, John; Eardley, Carolyn; Edwards, Malcolm; and Stableford, Brian, Eds.; Doubleday: New York; 1979, pp 440–441.

8. "The Court of Tartary", In *The Best from Fantasy and Science Fiction*; Davidson, Avram, Ed.; Doubleday: New York; 1965, p 79.

9. Caravan, T. P.; "Happy Solution", In *Other Worlds Science Stories*; January, 1952, p 93.

10. Caravan, T. P.; "Last Minute", In *Other Worlds Science Stories*; Oct., 1952, p 132.

11. Caravan, T. P.; "Dinosaur Day", In *Other Worlds Science Stories*; April, 1953, p 28.

12. Caravan, T. P.; "The Cold, Cold Grave", In *Other Worlds Science Stories*; May, 1953, p 97.

13. Parker, Helen N. *Biological Themes in Modern Science Fiction*; UMI Research Press: Ann Arbor, MI; 1984, p 60.

14. Malzberg, Barry N. *The Engines of the Night*; Bluejay Books, distributed by St. Martin's Press: New York, 1984, p 67.

★9★

On the Covers of Science Fiction Magazines

Jack H. Stocker

★One of the more striking features of the science fiction literature is that almost all of the older material appeared first in
the pulp magazines; even the more recent material was often
★ released initially in soft cover presentation. A great many SF
stories have never been offered in hard cover form. This emphasis
on magazine/soft cover presentation put a premium on the visual impact of
the covers of these books and magazines, designed to compete for the
attention of the browser. That the cover art often bore a flawed relationship
to the contents was a common occurrence. It also tended to be of mediocre
quality, poorly reproduced, lurid, and in vivid color, designed with cost and
attention-getting in mind. Certain publications, e.g., *Planet Stories* in the
late 1940s, were characterized by the regular cover appearance of some
scantily clad and well-endowed young woman, protected only by an apparent fishbowl helmet in high-vacuum outer space, clearly terrified, who was
being menaced by a drooling, multiply-tentacled (lustful? just hungry?)
monster. A well-muscled, similarly unprotected male brandishing a ray gun
was invariably speeding towards the pair. While the cover artist was usually
identified in the magazines, this was not always true in the earlier paperback
publications. There were, however, certain well established and much

admired artists in this earlier period: Frank R. Paul, Virgil Finlay, Hannes Bok, Chesley Bonstell, among others, whose work regularly graced the covers of the pulp magazines. Collections of their works expressly devoted to reproductions of SF cover art have appeared (see Appendix). The original works now command prices that would have astonished the artists who were paid very poorly for them.

Offered herewith are copies in their original colors of eight science fiction magazine covers selected, in part, for their (sometimes tenuous) chemical relevance. They illustrate general categories of SF magazine cover offerings.

★   ★   ★   ★

Cover 1. *Amazing Stories*, August 1926.

Some 70 years ago, the earliest SF magazines supplemented the limited availability of new material by offering reprints of SF stories written by distinguished authors. This cover indicates that the issue includes H. G. Wells's *War of the Worlds* along with fiction by Julian Huxley, the famous English biologist. Although Wells wrote the bulk of his notable science fiction in the last decade of the previous century, he was still writing in the genre until his death in 1946. He was very much alive at the time of this reprinting. It is an effective reminder of just how young the SF field really is.

Hugo Gernsback is identified on the cover as the magazine's editor. His writing was characterized by long dull stretches of scientific exposition/explanation; his messianic fervor to promote "science" is recognized and respected in naming the most prestigious SF award, the Hugo, after him.

Reprinted with permission of TSR, Inc.

Cover 2. ASTOUNDING Science Fiction, August 1942.

The story *Waldo* is an early novel by Robert Heinlein, destined to become perhaps the most highly respected author the field has yet produced. He was writing under the pseudonym of Anson MacDonald, derived from his wife's maiden name and used frequently to avoid recognition of his having authored more than one story in a magazine's single issue. The major figure in the novel, Waldo Farthingwaite Jones, is afflicted with myasthenia gravis; as a consequence he had but negligible muscular strength. He developed instrumentation that permitted him to magnify his arm movements such that he could not only marshal great strength but could control it with great precision at a distance. He named these devices after himself, calling them Waldos. Some years later these devices came into common use, particularly for the manipulation of highly radioactive materials from a safe distance. Enough of the scientists who actually made use of them apparently remembered the story and rewarded Heinlein's prescience (a very appropriate pun) by formally adopting his designation and calling their devices Waldos.

Yes, real scientists do read and remember science fiction. I recall a summer I spent as a researcher at the Oak Ridge National Laboratory four decades ago. When the regular supplies of popular magazines arrived at the town center's drug store, they would include a dozen of this and a dozen of that magazine but I distinctly remember seeing a five-foot-high stack of *Astounding Science Fiction* among them.

Cover 3. Astounding SCIENCE FICTION, December 1954.

A staple for artists doing science fiction covers was the presentation of what might be called Heavenly Viewpoints, characterized by precise, almost photographic-quality, representations, usually of planetary bodies, appropriately illuminated by our sun and viewed from very distant locations. It was considered critical that the representation be fully accurate. This genre was probably initiated and was most certainly exploited by the artist Chesley Bonstell. This cover is characteristic of his style. It can provide the viewer with a sense of the beauty and the immensity of the heavens and was used with great effectiveness in the movie 2001. Such representation is, if anything, even more widely practiced today by a number of artists, producing work of remarkable breadth and almost awesome photographic quality. One can stand almost anywhere in our solar system and, if illumination is right, snap a picture in any direction.

Mars seen from Phobos

Cover 4. Astounding SCIENCE FICTION, December 1953.

The featured story "Hide! Hide! Witch!" deals with the present and possibly growing anti-science sentiment in the world today. It extrapolates this hostility into a future in which all scientific research is effectively suppressed or subverted by the government and scientists are viewed as agents of Satan, hunted down and burned as witches. Just as there is a cliché of mad/bad/self-centered and irresponsible scientists, so is there a mirror image one of strongly sane/good/civic-minded and responsible scientists, a species much admired by hard-core science fiction fans. Many of the more thought-provoking SF stories deal with the ethics of a line of research that carries a significant element of risk and could yield results of a devastating nature (e.g., destroying all human life on the planet; recall *On the Beach*) or to be utilized for serious mischief (e.g., the study of relative ethnic intelligences for the Genome mapping project).

Science fiction readers tend to be a science-optimistic lot and the writers—reflecting in perhaps a majority of cases the next evolutionary stage of young SF readers—tend to produce upbeat stories. The cover story here is typical; the beleaguered scientist is "rescued" by the well-organized scientific community operating secretly underground where "good" science is carried out both responsibly and unfettered. Several characteristics of traditional SF fans are illustrated here: a general optimism, a trust in the results of science, a modest paranoia (they are out to get us!) and an assumption that we will have to fight back, i.e., rebel against the establishment. These kinds of stories reflect an admirable segment of the science fiction works; regrettably, they usually offer simplistic solutions, most frequently *via* some new technological breakthrough that is invoked to totally solve the problem.

Copyright ©1953 by Street & Smith Publications, Inc., reprinted with permission of Dell
Magazines, a division of Crosstown Publications.

Cover 5. Analog Science Fact – Fiction, July 1961.

Science fiction magazines traditionally offer a single science fact article per issue. These reports often differ from the accompanying unabashed fiction only in that they present the background facts in textual fashion and spend much of the article speculating on the potential consequences. The topics are usually selected, of course, to reflect recent dramatic developments that may have major repercussions in areas of interest to science fiction readers. The featured article, which appeared in July, 1961, considered a future mining of the asteroids by a portable power source and suggested it could be feasible by the year 1995. Obviously there would be some interesting chemistry attached to such an enterprise. The April 10, 1995 issue of *Chemical and Engineering News* carries on page 10, a report of the "first successful test of a complete, fully integrated solar dynamic space power system… in a space simulation vacuum". This correspondence of dates is rather striking!

Cover 6. Astounding SCIENCE FICTION, May 1953.

This issue featured a science-fact article on Interplanetary Industrial Design entitled "Space, Time and Education" which is reprinted in full elsewhere in this book. The cover offers a highly stylized, semi-abstract representation of a hypothetical planet (Arcturus), including its inhabitants, described in extensive detail in a brochure provided to the students in an M.I.T. course in Creative Engineering. The students were then asked to design appliances, conveyances, and power stations as well as other objects of interest to the Arcturians. The objects must be fully functional (as well as morally acceptable!). This type of cover with its stylized, star-like points of light is widely imitated and is almost a trademark for certain artists.

Cover 7. Astounding SCIENCE FICTION, October 1953.

Not only is the matter of sentience (a biochemical concern?) a primary SF theme, the interface between "human" and "machine" viewpoints continues to be a favorite topic. It is a widely held science fiction view that one of the most human drives is the satisfaction of curiosity. How and why do things work? Can machines share this quality? Does artificial life/artificial intelligence share this characteristic? Could robots display it? In more recent science fiction this question has been extended to computer program-derived "entities" presumably possessing humanlike qualities. The cover, by much-admired artist Kelly Freas, eloquently addresses this wish to understand.

October 1953 · 35 Cents

Astounding
SCIENCE FICTION

The Gulf Between BY TOM GODWIN

Cover 8. Galaxy Science Fiction, July 1953.

The SF author in search of a plot can often find inspiration from simply transporting the known and familiar to a galactic background. Issac Asimov's widely admired Foundation series took as its model the sequence of events leading to the fall of the Roman Empire and played it out on a field of galactic dimensions. Obviously such transformations need not be of such a magnitude; many delightful SF stories involve a much smaller scale and may involve reconfiguring the familiar and the absurd of everyday life. The magazine *Galaxy Science Fiction* featured for many years on the covers of its December issues a conventionally attired Santa with four arms, leading his space-protected reindeer on his Christmas errands, perhaps to a space station, perhaps waving, in passing, to an alien Santa in a sleigh pulled by lizard-like steeds or otherwise involved in traditional Christmas activities flavored with an alien component. This particular series appeared throughout much of the 1950s, created by the prolific and cherished artist Emshweiler who signed his drawings simply EMSH. The cover offered here is by this artist and is a personal favorite in that it reminds me of my teaching days many years ago at a state college in Mississippi. The state still had open-range laws in those days and more than once I had students apologize for missing a class; their car had hit a wandering cow on the highway. Now simply transport that event into interstellar space...

The New Media:
Television and the Movies

★ 10 ★

Where No One Has Gone Before

Chemistry in *Star Trek*

Natalie Foster

★ *Space: the final frontier. These are the voyages of the starship* Enterprise.

★ The 23rd century began with these words in the autumn of 1966. The original *Star Trek* series, first broadcast from 1966 to 1969, introduced television audiences to the universe of phasers, dilithium crystals, and warp drive. This series was followed by *The Next Generation*, which began filming in 1987 and aired its final episode seven seasons later in the spring of 1994. *Star Trek: The Next Generation* extends our view of the *Star Trek* universe from 80 years beyond that of the original series into the 24th century.

Although the two programs are first and foremost entertainment and are more stories about adventure than they are discourses about science, several shows have used similar scientific concepts as plot devices. Certainly in terms of space and time *Star Trek* transports us "where no one has gone before", but in terms of *chemistry*, by examining differences in the way the original series and *The Next Generation* treat the same topics, we can see where we have gone in science during the 25 Earth years that separate the two programs.

157

Basic Science: On Chemistry and Aging

The universal problem of growing old is confronted by the crews in both programs.

The Stories

In "The Deadly Years" (1967: Stardate 3478.2), an away team consisting of Captain Kirk, Dr. McCoy, Mr. Spock, Mr. Scott, Ensign Chekov, and Lt. Galloway visits a science station on the planet Gamma Hydra IV. Although the roster of inhabitants indicates that the head of the station is only 29 years old, he and all his colleagues appear extremely elderly. Once back aboard the *Enterprise*, the members of the away team—except for Mr. Chekov—begin to age at the rate of 30 years per day. Captain Kirk implores Dr. McCoy to do something. McCoy responds that he doesn't know what the problem is and suggests that, be it a virus, bacteria, or even evil spirits, he is determined to solve the medical mystery. Kirk points to Mr. Chekov and notes that he is the one member of the away party who has not fallen prey to the malady—he has not aged. Kirk wonders aloud about the reason for this anomaly and asks rhetorically if Chekov is immune simply because of his youth or if other factors like blood type, family characteristics, his glandular condition, or even his genetic make-up could be responsible. McCoy does not respond directly but grumblingly discounts the suggestion that Chekov's genes have anything to do with his resistance.

Mr. Spock soon identifies the cause of the premature and rapid aging: exposure to low levels of radiation from the tail of a comet through which the planet passed. A visiting endocrinologist on board, Dr. Janice Wallace, agrees to assist McCoy in finding a treatment and encourages Kirk by pointing out that her own line of research may prove promising in this situation. She is evaluating mixtures of carbohydrates for their ability to retard degenerative processes in plants. She further adds with great confidence that it is absolutely logical that they will find a solution to the premature aging because they have identified the problem and also know its progression.

The collapse and death of Lt. Galloway in sickbay spur a last ditch effort to understand why Chekov is not aging. Kirk and McCoy recall the events on the planet's surface. Kirk remembers that Chekov was terrified by the discovery of a dead body. Remembering the depth of Chekov's fright, McCoy catalogues the physical changes that accompany terror—rapid heart beat, shortness of breath, sweat, all the manifestations of a shot of adrenalin! He also recalls his lessons about the post-atomic age on earth and remembers that adrenalin was used to treat radiation sickness. The aged McCoy with the assistance of Dr. Wallace prepares adrenalin in the medical laboratory and administers it to all afflicted. The aging process is not only stopped but reversed, and the *Enterprise* warps off to its next mission.

The crew of *The Next Generation* encounters a very similar problem in "Unnatural Selection" (1989: Stardate 42494.8). The *Enterprise* responds to a distress signal from the U.S.S. *Lantree* only to find the entire, youthful crew dead from old age. The last port of call of the *Lantree* was the Darwin Genetic Research Station on Gagarin IV. Upon their arrival at Gagarin IV, Captain Picard and Dr. Pulaski of the *Enterprise* contact Dr. Kingsley of the Research Station. Upon hearing Dr. Pulaski's name, Dr. Kingsley asks her if she is the author of *Linear Models of Viral Propagation*. Pulaski is the author, and Kingsley promptly identifies that work as the standard in the field and expresses great relief at working with an expert.

Kingsley then describes what is happening at the Station, where the sole subject of the work is human genetics. She assures Picard that nothing has gotten out of control and that her most pressing concern is for the children at the Station, who are the result of years of genetic research. She wants Picard to evacuate the children to a safe haven aboard the *Enterprise*. Picard expresses amazement that there are children at the station, and Kingsley describes their rather nontraditional origin.

The scientists at the Darwin Station have engineered a group of children who represent an advanced human species. Captain Picard, fearing the children may be implicated somehow in the rapid aging of the normal humans to whom they have been exposed, requires positive proof that the children are harmless before he will honor Dr. Kingsley's request to evacuate them. Dr. Pulaski examines one child, transported to the *Enterprise* encased in styrolite (a transparent but impermeable containment material) and held in a force field, and declares that it may not even be possible for the child to contract a disease because his immune system is so highly advanced.

To comply with the Captain's wishes regarding the safety of the crew, Dr. Pulaski and Mr. Data, an android immune to any human pathogens, take the child outside the *Enterprise* in a shuttle craft to continue the examination by removing the styrolite. Immediately after freeing the child from the encasement, Dr. Pulaski experiences arthritic inflammation in her elbow and realizes that she has been afflicted and that the children are carriers of the malady; she begins to age rapidly. She requests that Data take her and the child to the Darwin Station.

Dr. Kingsley is incredulous at Dr. Pulaski's assertion that the disease was transmitted to her by the child. She describes the children's immune system as conferring upon them an aggressive immunity. The children produce antibodies capable of altering the genetic code of any virus they encounter. The antibodies actively find an airborne virus and attack it, not waiting for the virus to invade the body first.

Using the computer system and databases at the Darwin Station under Dr. Pulaski's direction, Data analyzes at a molecular level the interaction that Dr. Pulaski reasons occurred between the children at the Station and one of the crew members of the *Lantree* who had the Thalusian flu. The results of Data's analysis are sobering. He speculates that exposure of the children to the Thalusian flu from the *Lantree* crew member triggered their aggressive immune systems. However, after attacking the virus, the antibodies sought out the viral source and also attacked the human carriers, altering human genes in the process. Data shows in a computer graphic that he has compared two sections of DNA, one normal and one altered, and has determined where two molecules have been transposed. The site is in the gene responsible for aging. Pulaski, knowing that our DNA replicates itself, realizes that the error is irreversible, and Data confirms this assessment.

Dr. Pulaski and the others at the Station, all now very elderly, know that Data has sounded their death knell. She explains what has occurred to Captain Picard and offers her final log entry containing an assessment of the disease. Once normal humans are afflicted with the altered gene, they become carriers. The genetically engineered children on the Station are not carriers, however, they are the cause. The genes that are altered control aging. Although the new children may be an advanced species, they are lethal to normal humans, and Dr. Pulaski advises they must be kept in isolation forever on the Darwin Station, which in turn must be held under strict quarantine.

But all have not been idle on the *Enterprise*! Dr. Pulaski's colleagues have reasoned that they could adapt the special abilities of the transporter to filter out altered genes in a person's DNA and patch in corrected sequences. To make this possible—and it has never been attempted before—Commander Riker and Data must find a sample of Dr. Pulaski's DNA before she was exposed to the disease. Because she was only recently transferred to the *Enterprise*, they have no medical records and no transporter trace. In frustration, Riker shouts that they must check Pulaski's quarters for a live cell to provide a sample of her original DNA.

Riker and Data dash off to her room, where Data plucks a hair from Dr. Pulaski's hairbrush. He immediately identifies it as a hair with follicle intact and hence a source of live cells. Using the DNA from the follicle to generate a molecular matrix, they restore Dr. Pulaski to her correct age by filtration through the transporter. The other adults at the Darwin Station are similarly rejuvenated. The show closes with Dr. Pulaski's log entry: She acknowledges that no experiment is a failure, because even errors can lead to increased understanding. Achievements all have a price, however, and the crew of the *Lantree* paid for the newly acquired knowledge of the genetic code with their lives.

The Chemistry

"The Deadly Years" relied on radiation as the scientific driver of its plot. "Unnatural Selection" was concerned with DNA, genes, and genetic engineering. Those facts alone underscore a major difference between the scientific view of aging in the 1960s and the 1980s. It is also intriguing that the two stories highlight weighty fears on the part of the public in both decades about scientific pursuits: radiation as an invisible killer from nuclear weapons, and genetic engineering as a source of potentially lethal biological agents.

Regarding the scientific view of growing old in "The Deadly Years", the story illustrates accepted theory surprisingly well. The free radical hypothesis of aging, formally proposed by Denham Harman in 1956 (*1*), correlated gerontological changes with random destruction of cellular material similar to the type of damage associated with radiation exposure. The cross-linking of collagen, the peroxidation of lipids, and structural decay of membranes are all associated with aging and are all shown to be produced by reactions involving either free radicals or radiation. Mr. Spock's analysis pointed to radiation as the culprit and was scientifically up to date for the 1960s and in tune with the biochemical reasoning of the time. It was clear that radicals in the body damaged cells by compromising biochemical processes and that exposure to radiation produced radicals. Radiation as a fearful commodity was also certainly much in the minds of the general public in the aftermath of the build-your-own-bomb-shelter era. Hence both the understanding of aging and the fear of radiation are presented in this story.

An additional comment by Dr. McCoy harks back to an even earlier scientific theory of aging, Max Rubner's metabolic rate of living theory (*2*). Rubner proposed in the early 1900s that animals with high rates of metabolism were characteristically short-lived; animals had a fixed level of energy available to them, and once it had been expended, that was it. When Lt. Galloway collapses in sickbay, Dr. McCoy offers a succinct explanation directly from the pages of Rubner's theory: he states that she died because her metabolism caused her to age more quickly than the other afflicted crew members.

Although no one was seriously working on reversing the aging process in the 1960s, and aging was looked upon as something that had to be accepted and not manipulated, considerable effort of chemists was directed toward finding compounds capable of offering some protection from the onslaught of radiation. Many of the compounds studied were sulfur-containing species that were shown to provide protection by scavenging free radicals—by reacting more readily with damaging radicals than endogenous biomolecules did to form relatively stable intermediates. One of the more potent radiation-protective drugs, β-mercaptoethylamine, was actually observed to lengthen the mean life-span of mice, again providing a link between radiation, free radicals, and aging (3, 4). Other compounds like AET (2-aminoethylisothiouronium bromide hydrobromide) and cysteamine and its derivatives (5, 6) were discovered to be active agents that could minimize the effects of radiation when administered shortly before exposure. Most of these compounds, however, were judged too toxic for effective use as radiation protectives in humans (7).

An interesting note is that concerns about the possible role of free radicals in the induction of cancer have stimulated the further examination of such radiation-protective compounds as possible therapeutic aids for malignancy (8). Also, the suggestions that foods like broccoli contain cancer-preventing compounds echoes recommendations from the Army Quartermaster that eating broccoli might provide some protection from damage caused by radiation (9).

Treatment of individuals who had suffered whole-body radiation injury was and is a clinical problem of great concern. No mention has been found of the use of Dr. McCoy's remedy, adrenaline, either as a radiation-protective compound or as a treatment for radiation sickness. Adjuvant therapy using adrenocortical hormones was apparently evaluated after World War II (9) but found to be inappropriate (7). But in fairness, glandular extracts have a long history in fact (10, 11) and fiction (12, 13) as wonder drugs for a variety of ailments including senescence, so the selection of adrenaline in this context has both firm medical and literary roots.

Another chemical species now thought to contribute to the declining function of cells and tissues with aging is glucose. Dr. Wallace in "The Deadly Years" comments on experiments using carbohydrates to slow down degeneration in plants. Food chemists have recognized for decades that glucose alters proteins, but it may contribute to degradation, not retard it. A group of complicated nonenzymatic reactions known collectively as Maillard or browning reactions describe the chemistry between glucose and proteins that results in the browning of food during cooking. Recent studies (*14, 15*) have demonstrated that similar reactions may occur in vivo and cause the increased cross-linking of proteins that contributes to loss of elasticity in connective tissues as we age.

As far as aging is concerned in the 1980s and 1990s, "all roads lead to the gene" (*16*), and *The Next Generation* was right on target by ascribing changes associated with aging to alterations in the DNA of the subjects. Recently a specific genetic trait has been identified in simple life-forms that may be responsible for longer life (*17*). Some organisms may be genetically encoded to produce more superoxide dismutase, a compound that neutralizes free radicals, and may therefore age more slowly than others. Furthermore, scientists (*18, 19*) have actually identified a single gene known as AGE-1 that can be manipulated in a nematode to extend the lifetime of the worm by 70%. Free radicals still play an acknowledged role in the cellular destruction that characterizes the process of aging (*20*), but the plan for and the roots of this complex process are firmly lodged in the genetic makeup of each individual.

Data's observation that two molecules had been transposed in Dr. Pulaski's DNA (we'll allow that he must mean bases or nucleotides) certainly describes a gene that has been manipulated. Although considerations of free radicals damaging the integrity of genetic programming were known in the 1960s, it is amusing to hear Dr. McCoy grumble when genes are mentioned as a possible source of Mr. Chekov's resistance to the rapid aging afflicting the others. Modern science has shown that one can manipulate genes to extend life; manipulating them to shorten life by accelerating aging is also feasible. That the changes occur so rapidly during the episode is difficult to explain with current science, but we must agree to allow some poetic license to move the story line along.

A few other asides in "Unnatural Selection" are bound to warm the hearts of scientists. The first confrontation between Drs. Kingsley and Pulaski, in which Dr. Kingsley cites an early publication by Dr. Pulaski as "still the standard", reflects a version of the dream of all who publish: to be known for what we have written and to be remembered and referenced in a time of great need. And Data does not fall into a common trap by using a hair sample for a DNA analysis—he knows to hold out for a follicle containing live cells to provide the polynucleotide.

Finally, on a more sociological note, the idea that the immune system of the children produced an antibody that not only attacked an insulting virus but also aggressively destroyed the virus's host is an offshoot of their genetic engineering that even the denizens of the Darwin Station in the 24th century did not expect. This unexpected outcome strikes a cautionary tone and illustrates a real concern, especially within the public, about the possible dangers associated with genetic engineering in addition to an awareness of its potential benefits. The engineered children could not contract disease and hence were definitely improved models, but they also proved unfortunately lethal to their human predecessors, which is obviously no improvement as far as we are concerned.

A note of circumspection in the 1980s from Dr. Pulaski on this point stands in sharp contrast to the confidence of the 1960s mirrored in a statement by Dr. Wallace: that because we know the problem and the progress of the disease, all we have to do is find the correct tack to take in our research and logic dictates that a solution will be forthcoming. We could do it all in the 1960s—probably even manage the effects of radiation—and everything would be fine. Dr. Pulaski's statement has been chastened by events of the intervening years: no experiment is a failure and even errors move us forward. But all advancement costs something. We can learn from our mistakes, but we now know that both failure and success come at a price. In 25 years our knowledge of the chemistry of aging has vastly increased; in the same 25 years the bravado of the 1960s has been tempered by risk–benefit analysis.

Concerns about aging—delaying it, stopping it, reversing it—have been part of the human psyche for thousands of years. In both "The Deadly Years" and "Unnatural Selection", accepted scientific theories of aging current at the time the episodes were written are discernible in the plots, and our increased understanding of DNA and the chemical reactions involved in growing old are clearly mirrored in the presentation. The fears of the public are reflected as well, however, and form an important thread in the fabric of the dramas.

Applied Science: On the Transfer of Information by Spies

Advances in basic science drive improvements in the consumer world, and fundamental research on the chemistry of DNA and the genetic code has provided fertile ground for development. Events in *Star Trek* also illustrate this progression from the basic to the applied.

The Stories

In "Journey to Babel" in the original series (1967: Stardate 3842.3), the *Enterprise* is transporting delegates to a diplomatic conference when a fragment of a signal from an unknown source is detected at the same time an unidentified vessel begins to pace the *Enterprise*. One of the delegates, Thelev, is discovered to be the source of the signal and a spy. Thelev is a surgically altered Andorian with a radio transmitter implanted in one of his antennae. The radio is small, it is powerful, but it is just a radio, and the encoded messages to his comrades on the unidentified ship are simple radio transmissions.

In the 1960s, James Bond notwithstanding, a radio was probably the best technology available to transmit coded information. As David Kahn, the president of the American Cryptogram Association, wrote in 1966 (*21*): "Today so much secret information is contained in coded radio messages that cryptanalysts can intercept more intelligence than secret agents can steal."

Contrast this mode of sending a message with the technique used in *The Next Generation* in "The Drumhead" (1991: Stardate 44769.2). A Klingon exobiologist, Jedan, participating in a scientific exchange, is suspected of involvement in a breach of security. The warp drive of the *Enterprise* has been disabled in an apparent accident at the same time plans for the dilithium chamber have appeared in the hands of the Romulans, arch enemies of the Federation. Mr. Worf, Chief of Security, displays a small, hand-held device and offers the following testimony in a hearing with the accused spy.

Worf identifies the device as Jedan's hyposyringe with which he treats his Baltmazar syndrome. However, the device clearly has other uses. It can read data from Starfleet isolinear chips, which means Jedan can gain access to digitized information in computers. The device can then translate that information using a code and express the data in terms of sequences of amino acids that are then combined to form inert proteins. The proteins dissolve in the fluid in the syringe, from which they can then be injected into someone. The encoded information is transported in the carrier's blood. Nora Satie, a visiting investigator from Starfleet, realizes immediately the magnitude of Worf's conjecture: the living body becomes a tool for the transportation of secret files. Worf states that the hyposyringe really only has one function—to translate digitized data from a computer into biological sequences.

The Chemistry

The idea that an encoded message can be transmitted as a protein is an intriguing application of the techniques for manipulating peptides developed in the quarter of a century separating the shows. The concept of coding information in biochemical sequences has its origin in the discovery of the double helical structure of DNA and the deciphering of the genetic code, major scientific breakthroughs of the 1950s and 1960s. Francis H. C. Crick, one of the scientists who uncloaked these mysteries, presaged the application of the amino acid code used in "The Drumhead" by pointing out in the 1960s that a protein was like a long sentence written in a language that has 20 letters (22). Most recently the magazine *Forbes* suggested (23) that "in the Nineties, the new technology theme is going to be codes—the creation and decryption of biochemical and electronic coding systems." The technological advances predicted in *Forbes* have borne fruit in the 24th century for the Klingons.

In addition, the development of peptides that do not trigger an immune response or that can withstand the degradative enzymes in the body is currently a major thrust in drug design (24). The problem of fishing Jedan's unique message-containing protein out of the carrier's blood is a daunting one (we must allow that separation science will surely progress by the 24th century), but the idea of preparing a protein resistant to the body's defense mechanisms is well established, and the actual encodement of data in biochemical sequences is readily conceivable.

Designer proteins are not the only noteworthy science in "The Drumhead". In describing their analysis of the exploded dilithium chamber, other members of the crew make it clear that they have all the equipment for modern analytical work that any chemist might want. Data describes analyzing the dilithium chamber using microtomography, and La Forge mentions using mass spectrometry to screen the chemical content of the blast residue. Tomographic imaging of solid fragments from explosions, coupled with mass spectrometry of volatile residues, is a powerful tool in the arsenal of modern forensic scientists seeking the causes of explosions and reconstructing blast patterns. The *Enterprise* is well equipped for its mission of scientific exploration.

From coded transmissions sent via miniaturized radios to messages carried by encoded proteins in vivo is a mighty leap for espionage. Although our Earth-based science cannot quite accomplish the necessary tasks to make the work of Jedan's hyposyringe a reality now, on the basis of what we have accomplished in protein manipulations since 1960, the chemistry is conceivable, and accepting the premise does not require a major leap of faith today.

Unsolved Problems: On Viral Infections and Alcohol

Problems we have not yet conquered still plague the occupants of the *Enterprise* in both centuries.

The Stories

In "The Naked Time" (1966: Stardate 1704.2), the *Enterprise* is ordered to pick up the staff at a remote planetary outpost and simultaneously observe the gravitational shifts associated with the disintegration of the planet. However, when the *Enterprise* arrives, everyone at the outpost is dead, and the circumstances are extraordinary: Life support at the post has been turned off, one woman has been strangled, and a man died while showering fully clothed. Mr. Scott informs the Captain with familiar words that, not only is the cause of these events unknown, they have not dealt with anything like it before.

When Dr. McCoy comments that the cause is definitely not drugs or intoxication because bioanalysis reported on the tapes has proved that conclusively, Mr. Spock, that eloquent spokesman for caution in interpreting data, gives us all an important reminder. He chides McCoy by reminding him that instruments only detect the events they are designed to detect; space presents us with an infinite variety of unknowns.

Soon individual members of the crew behave as if intoxicated; inhibitions are released, and behavior becomes irrational. The malady spreads. Finally Dr. McCoy discovers the cause—a bizarre virus. The virus is water, changed into a complex chain of molecules by the gravitational shifts on the planet. This explains why it was both impossible to detect and so easy to transport. The agent passes from person to person by direct contact through perspiration. Once it infected an individual, the symptoms are like intoxication only much more pronounced. People lose self-control, judgment, and inhibitions.

Anyone who remembers the "polywater" controversy in the 1970s (25, 26), long after this episode was consigned to reruns, might speculate at this point about life imitating art. Polywater, or anomalous water, was the name given to an alleged polymeric form of water. Evidence for it was discovered in small capillary tubes, and it was proposed to be the form of water present on Venus. Additionally, it was described as a potential force that, if loosed to the environment, might serve as a condensing point and turn all water into polywater and destroy all life (25). It was also touted as a possible antifreeze or a material to desalt water (26). Its existence has never been verified.

A strikingly similar scenario unfolds in "The Naked Now" (1987: Stardate 41209.2) for the heroes of *The Next Generation*. The *Enterprise* contacts the crew of the science vessel *Tsiolkovsky*, who are observing the collapse of a star from a red giant into a white dwarf. However, when the *Enterprise* arrives at the scene, the *Tsiolkovsky*'s crew is dead. Sent to investigate, the boarding party reports that it appears as though a wild party has taken place. When asked by Picard for a theory. Dr. Crusher allows that it could be any number of things including madness, mass hysteria, or delusional episodes.

Upon the return of the boarding party, the crew aboard the *Enterprise* begins to experience apparent intoxication with concomitant release of inhibitions and irrational behavior. The empathic Counselor Troi consults with Dr. Crusher about an affected crew member, describing that she senses mostly confusion from him of the sort that results from consuming too much alcohol. Crusher responds that intoxication would have shown up in her medical evaluation, as well as any other signs of drugs or other external agents. They could have used Mr. Spock's wisdom here about what things instruments can detect.

Commander Riker recalls a historical account of a similar occurrence and, with the help of Data, searches the computer's historical file and discovers amazing parallels to their situation and events on the old *Enterprise*. He discovers in the historical record that changes in gravitational fields of the same magnitude caused similar results. Water became polymeric and acted on the brain like alcohol, only with much more dramatic effect. He and Data also find the formula developed by Dr. McCoy that was successful in treating the virus 80 years ago. They transmit the formula to sickbay immediately, but unfortunately, the old formula does not work because the virus has mutated. Dr. Crusher must isolate the virus and prepare a new antiviral remedy, which she does in time to save the ship from the exploding star.

The Science

Viral mutations were certainly known in the 1960s. We now may know much more about viruses, but we still face the problem of their mutation (27) in providing long-term effectiveness in vaccines or treatments, as anyone who contracts the flu after being immunized realizes. Perhaps we can draw comfort from the fact that the battle between viruses and humans is still going on in the 23rd and 24th centuries, and that we have not yet lost.

The treatment of the effects of alcohol on human behavior is also not scientifically enlightened in either program and is handled much the same way in both. Violent behavior is noted among the crew at the outpost in "The Naked Time" and aboard the *Tsiolkovsky* in "The Naked Now". Although members of both *Enterprise* crews lose control, the original crew seems more violent than their colleagues 80 years hence, but that is a characteristic difference between the two shows that is not just confined to these episodes. The correlation between consumption of alcohol and violence is widely accepted anecdotally in addition to being statistically valid (28). The physicians in both situations talk of people losing capacity for judgment and self-control. Inhibitions about sexual activity among crew members are released, and even the stoic Mr. Spock and the android Data are affected and involved in a "piece of the action". The folk wisdom about alcohol hasn't changed much in 25 years, and neither show transcends that level of approach to the scenario presented.

The Ultimate Optimism

Twenty-five years have had their impact on *Star Trek* in many ways: no *one* now goes where no *man* had gone before; the Klingons are now our allies; we do seem to negotiate more and fight less. And as the episodes cited herein demonstrate, 300-plus years in the future we still grow old, practice espionage and diplomacy simultaneously, and get the flu. But the ultimate optimism of *Star Trek* is that we still do exist, we care for each other, we strive to lead ethical lives, and we write bawdy limericks. We continue to learn, too, and the years between the original series and *The Next Generation* have seen a remarkable expansion of chemical knowledge here on Earth, which has occasionally been refreshingly reflected in new approaches to familiar problems encountered during voyages to the final frontier.

Acknowledgments

This article is taken in part from "Science and the Final Frontier: Chemistry and *Star Trek*", by Natalie Foster, originally appearing in *Chemistry & Industry*, Volume 24, December 21, 1992, pp 947–949.

I acknowledge Tim Lynch, Chris Harmon, and the other regular posters to rec.arts.startrek for their frequently stimulating discussions.

- "The Deadly Years", by David P. Harmon, originally telecast December 8, 1967.

- "Journey to Babel", by D. C. Fontana, originally telecast November 11, 1967.

- "The Naked Time", by John D. F. Black, originally telecast September 22, 1966.

- "Unnatural Selection", by John Mason and Mike Gray, originally telecast January 28, 1989.

- "The Drumhead", by Jeri Taylor, originally telecast April 27, 1991.

- "The Naked Now", Johnny Dawkins, Ed., originally telecast October 3, 1987.

References

1. Harman, D., "Aging: A Theory Based on Free Radical and Radiation Chemistry", *J. Gerontol.* **1956**, *11(3)*, 298–300.
2. Rubner, M., "Energy and Matter in the Living Organism", *Chem. Ztg.* **1907**, *32*, 930–931.
3. Bacq, Z. M.; Dechamps, G.; Fischer, P.; Herve, A.; Le Bihan, H.; Lecompte, J.; Piorotte, M.; Rayet, P., "Protection Against X-rays and Therapy of Radiation Sickness with β-Mercaptoethylamine", *Science (Washington, D.C.)* **1953**, *117*, 633–636.
4. Dacquisto, M. P.; Benson, S. M., "Role of Mercaptoethylamine in Repeated Monthly Exposure to Gamma-Radiation in Mice", *Nature (London)* **1962**, *195*, 116–117.
5. Maisin, J. R.; Doherty, D. G., "Comparative Chemical Protection to the Intestinal and Hematopoietic Systems of Whole-Body X-Irradiated Mice", *Radiat. Res.* **1963**, *19*, 474–484.
6. Bekkum, D. W.; Nieuwerkerk, H. T. M., "The Radioprotective Action of a Number of Cysteamine Derivatives and Related Compounds", *Int. J. Radiat. Biol.* **1963**, *7*, 473–479.
7. *The Treatment of Radiation Injury*; National Academy of Sciences and National Research Council Pub. 1134, National Academy of Sciences: Washington, DC, 1963, p 11.
8. *Radiation Sensitizers*; Brady, L. W., Ed.; Masson Publishers USA: New York, 1980.
9. *Exposure of Man to Radiation in Nuclear Warfare*; Rust, J. H.; McWissen, D. J., Eds.; Elsevier: New York, 1963, pp 136–139.
10. Waller, T. E. *Theory and Practice of Thyroid Therapy*; John Bates and Son and Danielsson, Ltd.: London, 1914, preface.

11. Cooper, D. S., "Thyroid Hormone Treatment: New Insights into an Old Therapy", *J. Am. Med. Assoc.* **1989**, *261*, 2694–2695.
12. Doyle, A. C. "The Adventure of the Creeping Man", in *The Complete Sherlock Holmes*; Garden City Publishing Company: New York, 1938, pp 1261–1276.
13. Sayers, D. L. *The Unpleasantness at the Bellona Club*; Avon Books: New York, 1963.
14. Vlassara, H.; Brownlee, M.; Cerami, A., "Nonenzymatic Glycosylation: Role in the Pathogenesis of Diabetic Complications", *Clin. Chem.* **1986**, *32*, B37–41.
15. Wolff, S. P.; Jiang, Z. Y.; Hunt, J. V., "Protein Glycation and Oxidative Stress in Diabetes Mellitus and Aging", *Free Radical Biol. Med.* **1991**, *10(5)*, 339–352.
16. Begley, S.; Hager, M.; Murr, A., "The Search for the Fountain of Youth", *Newsweek*, March 5, 1990, pp 44–48.
17. Munkres, K. D., "Genetic Coregulation of Longevity and Antioxienzymes in *Neurospora crassa*", *Free Rad. Biol. Med.* **1990**, *8(4)*, 355–361.
18. Gupta, A.; Hasan, M.; Chander, R.; Kapoor, N. K., "Age-Related Elevation of Lipid Peroxidation Products: Diminution of Superoxide Dismutase Activity in the Central Nervous System of Rats", *Gerontology* **1991**, *37(6)*, 305–309.
19. Friedman, D. B.; Johnson, T. E., "A Mutation in the Age-1 Gene in *Caenorhabditis elegans* Lengthens Life and Reduces Hermaphrodite Fertility", *Genetics* **1988**, *118*, 75–86.
20. Johnson, T. E., "Increased Life-Span of Age-1 Mutants in *Caenorhabditis elegans* and Lower Gompertz Rate of Aging", *Science (Washington, D.C.)* **1990**, *249*, 908–912.
21. Kahn, D. "Modern Cryptology", *Sci. Amer.* **1966**, *215(1)*, 38–46.
22. Crick, F. H. C. "The Genetic Code", *Sci. Amer.* **1962**, *207(4)*, 66–74.
23. Gianturco, M. "Theme for the Nineties", *Forbes* January 8, 1990, p 304.
24. Krieger, J. "Protein Drug Delivery System Nears Approval", *Chem. Eng. News* **1990**, *68(3)*, 38.
25. "Polywater Controversy Boils Over", *Chem. Eng. News* **1970**, *48(27)*, 7–8.
26. "Polywater Existence Still Unsettled", *Chem. Eng. News* **1970**, *48(29)*, 29–35.
27. Henig, R. M., "Flu Pandemic: A Once and Future Menace", *The New York Times Magazine*, November 29, 1992, pp 28 ff.
28. "Alcohol and Violence", *The Lancet* **1990**, *336*, 1223–1224.

* 11 *

Chemistry in TV Science Fiction

Star Trek and Dr. Who

*Penny A. Chaloner**

★I first became interested in the relationship between TV science fiction and real science when watching reruns of *Star Trek*, the original series (TOS), episode "The Naked Time" *(1)* and the *Star Trek: The Next Generation* (TNG) episode "The Naked Now" *(2)*. In both of these episodes the crew become contaminated with a form of water that acts as an intoxicant and are cured just in time to save the *Enterprise* from disaster. I wondered about a connection with the great polywater saga of the late 1960s—was art imitating life or life imitating art? In fact there is no apparent connection, at least with the earlier episode. The initial papers on polywater (3) did not appear until after "The Naked Time" was transmitted, and the idea may owe more to the Kurt Vonnegut novel *Cat's Cradle* (4), which introduces

---

*This chapter was originally written with short quotations illustrating the points made. All were carefully and expressly referenced. However, Paramount refused permission for any quotation whatsoever. It seems extremely sad that Paramount is unable to distinguish between scholarly work from which the authors derive no pecuniary benefit and large-scale commercial attempts to infringe copyright for profit.

173

us to ice-nine. This was a polymorph of ice that melted well above room temperature, and, once formed, triggered the conversion of all the world's water to that polymorph, with obviously disastrous consequences. This book predates *Star Trek*.

However, my initial contention remains. Science fiction on TV is extremely important in influencing the public perception of science. In a typical week, the first run of TNG episodes had an audience of 17 million, and both *Star Trek: Deep Space Nine* and *Star Trek: Voyager* have attracted huge advertising revenue. Despite poor scheduling, the United Kingdom terrestrial audience for TNG was more than 5 million for new episodes (even though they have been previously shown on satellite and had been available for purchase on video for more than a year), and 3–4 million for reruns of the original series, made before a large proportion of its audience was born. All eight movies have been huge box office hits. The series have been translated into 47 languages, and it is estimated that episodes are screened 200 times a day on TV stations in the United States. These statistics should be compared with the tiny audiences for "serious" science programs such as *Nova* and *Horizon*.

It was no coincidence that the first U.S. space shuttle was named *Enterprise*; in 1976, NASA received more than 400,000 letters suggesting this name from *Star Trek* fans. The Bank of Scotland has issued a *Star Trek* credit card, and *Star Trek* coins are legal tender in Liberia. Revenue from merchandising has topped $1 billion. On the Internet, 10,000 messages a month appear about *Star Trek*, and it has had 1299 mentions in U.K. broadsheet newspapers in the past 10 years.

Dr. Who ran for almost 30 years and was sold to 60 countries, with a regular viewing audience of 110 million (5). Although no new episodes had then been made for 6 years (other than a brief charity special, and two new adventures on radio), in 1995, British Telecom chose the image of a Dalek to advertise Phoneday, when all U.K. area codes were to change. The slogan was "It's one to remember", as most of the new codes had inserted the digit 1. The word "Dalek" has made it into the Oxford English Dictionary.

Why are the series so popular? Why did these rather than others catch the public imagination? One theory is that the series present an optimistic view of the future in a genre that has been dominated by dystopic scenarios. Maybe we just like to see the good guys win.

The era with which I want to deal begins with the first U.K. transmission of *Dr. Who* in November 1963. Science fiction series existed before this time; notable examples were *The Outer Limits, The Twilight Zone, Flash Gordon,* and *Quatermass,* but only in the last two of these was there a regular cast, and only *Flash Gordon* could accurately be assigned to the space opera genre, which has since become so popular. Since then there have been many other significant and popular series, including *Lost in Space, Land of the Giants, Thunderbirds* (which recently enjoyed a very popular revival in the United Kingdom), *Buck Rogers, Battlestar Galactica, Space 1999, Blake's Seven, Babylon 5,* and *Space Precinct.*

In this chapter I want to pose, and I hope to answer, three questions:

- Do these series get their chemistry correct?

- How does chemistry fare relative to other science disciplines in terms of the sympathy with which it is portrayed and the time that is spent on it?

- Do these series reflect contemporary thinking and concerns about science?

Is the Chemistry Correct?

Dr. Who ran almost every season between 1963 and 1989, and there have been many reruns. A new TV movie starring Paul McGann as the Doctor was shown in May 1996, and the interest of fans has been sustained by novels of the "New Adventures" and the "Missing Adventures", and two radio adaptations (6). The Doctor says on several occasions that he is not a doctor of medicine (although he does have a medical degree). He seems closest to being a physicist, given his expertise in time travel. The sonic screwdriver is probably misleading; it is closer to being a magician's wand than a real scientific tool. In general, though not invariably, he gets his science right. This outcome reflects the high quality of writing in the series and the fact that a number of professional scientists have been involved (for example, Dr. Kit Pedler in the later episodes of William Hartnell's Doctor).

In the episode "The Sensorites" (7), from the Hartnell era, the emission lines for various elements are identified in "spectrographs", which is reasonable if this is viewed as atomic absorption spectroscopy. The quoted melting points for iron and molybdenum are very close to correct, and the only howler is that the Doctor treats Ian's atropine poisoning with caffeine citrate; I doubt that this would contribute much to his recovery.

In "Planet of the Giants" (8), the Doctor and his companions are accidentally miniaturized, and the *Tardis* lands in an English country garden. They find the formula of an insecticide, DN-6, and Barbara is accidentally poisoned by it. The formula shown looks authentic, and the comments are sensible, given that organophosphates have found wide use as insecticides.

One of the first occasions on which the Doctor does any practical chemistry is in the story "The Krotons" (9), an episode in which Patrick Troughton plays the Doctor.

At the start of the story, magnesium silicate is not quite correctly identified by Zoe Herriot, one of the smarter companions, as mica. The Krotons are a life system based on tellurium, which must be considered to be fairly far-fetched. The Doctor and Zoe talk about tellurium as they wander through the wastelands; she is aware of the appalling smell of hydrogen telluride and correctly identifies the atomic weight and number of the element.

Later the Doctor uses native sulfur to prepare sulfuric acid, and the Krotons, who are 80% tellurium, dissolve in it. In "The Dominators" (10), he makes bombs from the medical kit.

An episode in which the science is rather poor is "The Seeds of Death" (11). A deadly foam arrives on Earth as a prelude to an invasion of the ice warriors from Mars. The foam removes oxygen from the air. It contains molecules of five atoms that are not affected by mineral acids but are destroyed by water. "The Claws of Axos" (12) also contains some poorly thought-out science. Axonite is said to be a "thinking molecule" that can reprogram and recreate. The Doctor accelerates it beyond the speed of light! One comment, however, must appeal to anyone who lived in the United Kingdom in the early 1970s; the Doctor advises his listeners to take the recommended precautions against a nuclear attack—"sticky [Scotch] tape to prevent the windows from shattering."

In "Planet of the Daleks" (13), we encounter an allotrope of ice that never becomes solid and forms the heart of the planet Spiridon. The Doctor uses this to freeze the Dalek army. This seems a fairly obvious reference to Vonnegut's ice-nine. In "Revenge of the Cybermen" (*14*), both the Doctor and his companions show that they know about the properties of gold; Harry knows that it is soft enough that battering his gold fetters with a rock will probably break them, and the Doctor explains that gold is fatal to the Cybermen, because it gold-plates their breathing apparatus and, being relatively inert, suffocates them.

In the story "The Hand of Fear" (*15*), the Doctor and Sarah-Jane meet a silicon life-form, Eldrad. The noncarbon life-form is a recurrent motif that we will meet again, but the science involved is barely discussed here. In the later episode, "The Stones of Blood" (*16*), the Ogri are also silicon-based, globulin-deficient (why should a silicon-based life-form need globulin?) life-forms, who seem to live on blood.

In the final episode of Tom Baker's Doctor, "Logopolis" (*17*), there are several discussions about the second law of thermodynamics. The statements are reasonably sound, correctly noting that in a closed system, entropy must increase. On Logopolis, the Doctor, Nyssa, and the Monitor discuss the nature of the Universe as a closed or open system: The solution to the collapse of the Universe that they predict, the creation of voids into other parallel universes, is rather fanciful, but the conceptual discussion of closed and open systems is quite sophisticated.

In "Time and the Rani" (*18*) the Doctor correctly uses the acronyms PHB (polyhydroxybutyrate) and PES (polyether sulfone) for plastic samples. He accurately identifies PHB as biodegradable and PES as petroleum-based—a good science advisor here obviously!

The original *Star Trek* series contains many dreadful chemical howlers. The *Enterprise* is powered by a matter–antimatter reaction that is controlled by using lithium ("Where No Man Has Gone Before") (*19*) or dilithium crystals (later episodes). Lithium–lithium bonds must be considered somewhat rare; the crystals are shown in "Elaan of Troyius" (*20*), and they have the appearance of quartz. The conditions to sustain life seem to be poorly understood. In "The Galileo Seven" (*21*), the atmosphere of the planet on which Spock and his party crash is described as having 70-mm pressure of oxygen and 140-mm nitrogen, which is described as quite reasonable. In "Space Seed" (*22*) the captain survives 10-mm pressure in a medical decompression chamber!

In "Arena" (*23*) Captain Kirk shows that he has not forgotten his general chemistry course. He identifies potassium nitrate (by its taste!)

and mixes it with sulfur and carbon to make crude gunpowder to fire diamond projectiles at his enemy, the Gorn. Kirk also professes belief in the possibility of a silicon life-form in "The Devil in the Dark" (24). Scientifically this is a poor episode; it simply picks up the alternative element life-form motif, which has been used with more conviction elsewhere. McCoy pours further scorn on the idea that a silicon-based life-form could exist in an oxygen atmosphere, though he does not make his reasons clear. Eventually he repairs the damaged Horta with thermal concrete, and the Horta's eggs, which appear to be silicon spheres, hatch successfully. No distinction is made between silicon and silica.

In the episode "Patterns of Force" (25), Spock improvises some quite neat chemical physics. He uses "rubindium" crystals from a transponder to construct a primitive laser to burn through the lock of their jail in the Nazi regime on Ekos. Spock remarks that "ancient" lasers achieved the necessary excitation even though they used crude natural crystals. The light from the laser that he constructs looks rather like that from the one that, less than 30 years later, I now carry in my pocket to use as a pointer in lectures.

In "The Omega Glory" (26), McCoy displays some serious ignorance about the composition of the human body. His analysis of the powder they have found is that it contains 35% potassium, 18% carbon, 1% phosphorus, and 1.5% calcium, these being the residue when a human body is dehydrated! It would seem that not only are his chemistry and biology at fault, but he is not too good at arithmetic either (his estimate that 96% of our bodies is water would have the crew weighing 75–100 pounds). Also, 96% water in the human body is a substantial overestimate, and what was left looked very much like rock salt.

An early episode of Star Trek: The Next Generation, "Home Soil" (27), returns to the theme of a noncarbon life-form, this time with rather more success. The inorganic life-form on Velara III is being destroyed by the terraforming team. It is discovered by Data and Geordi and beamed to sick bay. There it starts to grow and develop, and eventually it communicates with the Enterprise crew. As they analyze it with the help of the computer, most of the science they discuss is correct, or nearly so. Data identifies silicon and germanium as being part of transistors, and he and Geordi are essentially correct in their description of the functions of gallium arsenide and cadmium selenide sulfide.

When the "microbrain" finally achieves direct communication with the crew, it addresses them as "ugly bags of mostly water". Data gives a figure of 90% for the water content of the human body, still somewhat high. Both this and the story "Evolution" (28), in which Wesley breeds brighter nanites, must be regarded as scientifically credible versions of the noncarbon life-form story.

The episode "The Royale" (29) is generally considered to be a weak one from a dramatic point of view, and there is also a terrible scientific howler. Geordi describes the planet below them as having an atmosphere of nitrogen and methane, with some liquid neon. He records the temperature on the surface as –291 degrees Celsius. Doubtless science will have undergone many changes by the 24th century, but a change in the value of absolute zero seems one of the less likely, and neon has a much lower boiling point than methane or ammonia. Another avoidable error appears in "Evolution" (28). The angry nanites flood the bridge with nitrogen oxide, which is described as N_2O. Unfortunately this looks like fog and leaves the bridge crew choking. It's hard to believe that no one around the set had ever had gas at the dentist, even if they had never taken a chemistry class.

The technique of DNA fingerprinting for identifying an individual or establishing a close family relationship is used in several TNG episodes. In "The Vengeance Factor" (30), Yuta of the clan Trelesta carries a genetically engineered virus that is active only against certain DNA types, those of her blood-feud enemies, the clan Lornak. To her it is quite harmless. DNA typing is used to establish the identity of Jono as Jeremiah Rossa in "Suddenly Human" (31) and Ishara Yar as Tasha's sister in "Legacy" (32). In "Data's Day" (33) the organic residue on the transporter pad does not correspond to T'pel's DNA, so Data is able to conclude that she is not dead, as they had been intended to believe. Genetic engineering is a feature of "Unnatural Selection" (34), and authentic DNA from one of Dr. Pulaski's hair follicles is used in the transporter "de-aging" process. Storage of information in the form of memory RNA is used by the ship piloted by John Doe in "Transfigurations" (35); Data and Geordi eventually decode this into a navigational log. A related biochemical device is used for information storage by a Klingon spy in "The Drumhead" (36). However, in later episodes there is some terrible biology ["The Chase" (37), "Sub Rosa" (38), "Genesis" (39)] and some truly awful technobabble. In

"Cost of Living" (*40*), Geordi says that elements in the replicator have been degraded into simpler *molecular* structures. Beverley, in Schisms" (*41*) gives Riker a hot toddy. She says that the lactose contains amino acids that are activated by heat to make it a sedative.

In the episode "Legacy" (*32*), Data shows that he has not forgotten his basic chemistry courses. He suggests that Dr. Crusher should be able to remove the proximity detector and air-sensitive microexplosive from under Ishara's skin by carrying out the operation under an atmosphere of xenon. The idea is reasonable and indeed works well. His chemistry looks a lot less impressive in "Night Terrors" (*42*). In this story the *Enterprise* is trapped in a Tyken's rift, and the crew are subject to a unique chemical imbalance caused by dream deprivation. Troi is the only one who is dreaming, and in her dream she is receiving a telepathic message from the crew of another ship, which refers to "eyes in the dark". Eventually she concludes that the other ship wants them to release something into the rift, to create an explosion. Troi and Data review the "elements" they have available to them on board. As they appear on the monitor, it is clear that these are mainly compounds. She identifies hydrogen with one electron circling one proton, but the picture looks much more like a hydrogen molecule, as indeed does her dream picture. A neatly contemporary touch is provided in "Firstborn" (*43*). Alexander has made models of fullerenes in his chemistry class; unfortunately he then fills them with water to throw at his friends. In "Time's Arrow" (*44*), C_{70} is a component of Data's communications badge.

The record of *Star Trek: Deep Space Nine* (DS9) contains little chemistry, and that little is poorly done. One of the worst howlers in the series comes in "A Man Alone" (*45*), when complex proteins are described as breaking down into DNA fragments! In "The Alternate" (*46*), the landing party describe a material they find as silicate, but vegetative, and showing distinct evidence of life. It looks rather like a pile of iron filings. Odo's shapeshifting always presents a scientific problem; he appears to be able to change his mass, as he has been disguised as a rat, a chair, a glass, etc. In "Broken Link (*47*), when he becomes ill, his mass and density fluctuate.

There is, however, quite a neat idea in "The Abandoned" (*48*). The Jem'Hadaar baby grows up very fast, but has been genetically engineered to be dependent on an enzyme, which Bashir has to administer. This situation is reminiscent of *Jurassic Park* (49), where the dinosaurs have been genetically engineered with a metabolic defect, which means that they cannot synthesize lysine and must have it in their diet. The theme recurs in "Hippocratic Oath" (50), when Bashir almost succeeds in synthesizing the "ketresel white", and "To The Death" (51), when we discover that "the white" is all the Jem'Hadaar need for nutrition and that they neither eat nor sleep.

Perhaps it is still too early to judge whether *Star Trek: Voyager* (STV) will get its science right. The story with the most obviously chemical theme in the first season was "Emanations" (52), in which the crew discover a new element, with unusual origins and properties. It "emanates" from decomposing alien bodies as part of their natural decomposition process. Janeway tells us that the Federation currently knows of 246 elements. The new element has a mass of 550, but is stable. Although this element does not seem very likely, the technobabble is not too grating.

In "Learning Curve" (53), Tuvok survives an environment in which the temperature rises to 360 K, 87 °C, which seems high even for a Vulcan. Rust is correctly identified as iron oxide in "The 37s" (54), but gasoline is said to consist of benzene, ethylene, and acetylene. Not only does this seem an unduly volatile mixture, but it appears that even in the 25th century, there are those who have not caught up with IUPAC nomenclature.

In "Cold Fire (55), Kes heats up a liquid telekinetically by making the molecules move faster, a reasonable way to do it if you believe in telekinesis. In "Basics—Part 1" (56), Harry Kim detects nitrogen tetroxide in the cabin of the shuttlecraft of the Kazon Nystrom spy. Presumably, he means dinitrogen tetroxide. If we assume a normal ambient temperature, this would be in the gas phase (boiling point, 21 °C), and there would be considerable dissociation to NO_2. He correctly realizes that the gas would be toxic and that the Kazon would not survive in it for long.

Terrible biology has been evident in several episodes of *Star Trek: Voyager*. In "Threshold" (57), Tom Paris starts to mutate after his transwarp flight; he becomes allergic to the water in his coffee and can no longer breathe oxygen. Treatment involves putting him in an atmosphere of 80% nitrogen and 20% "acid dichloride" and dosing him with radiation up to 100 rads per second! In the episode "Tuvix" (58), Tuvok and Neelix are fused by the transporter; apparently lysosomal enzymes in some alien orchids led to symbiogenesis (and the breaking of the laws of the conservation of matter).

How Are Scientists Portrayed?

Turning to the way in which scientists in general, and chemists in particular, are portrayed, the early episodes of *Dr. Who* look promising. One of the first companions was Ian Chesterton, Susan's science teacher. Unfortunately his main function is in an "action man" rôle; he is brave and resourceful, but the intellectual part is left to the Doctor. Zoe (the Doctor is almost as clever as I am) Herriott is a good image of a scientist, but her specialization is computing. Liz Shaw, who arrives in 1970, is a distinguished Cambridge scientist with degrees in physics and medicine, but is still addressed throughout as Miss Shaw. One suspects that the Brigadier would really prefer that she made the tea. Later companions with a scientific background include Adric, a mathematician, and Nyssa, who is essentially a biochemist. Ace is one of the most convincing of the companions; although she failed her chemistry exams, she seems to be adept at practical work and makes some excellent bombs. She was suspended from school for blowing up the art room. Thus, among the companions, who are all positive figures, there are a few scientists, but chemistry is not prominent. The Doctor's archenemy, the Master, is clearly an excellent scientist, but does not do much chemistry. The Rani, another renegade Time Lord, is described as a brilliant chemist, but the Doctor considers that she sees humans as soulless heaps of chemicals, which he says is typical of a certain type of scientist. The Doctor defeats her in all their encounters, but like the Master, she lives to fight another day.

Other scientists portrayed in the series are a much more mixed bunch. In "Planet of the Giants"(8), the chemist who made DN-6 is portrayed as entirely amoral and profit-oriented. Dr. Tyler, the astrophysicist in "The Three Doctors" (59) is a good strong character; he is both interesting and interested—and also brave, as he is the first to offer to pass through the singularity to return to Earth. Both "The Green Death" (60) and "Genesis of the Daleks" (61) tell stories in which two groups of scientists, one "good" and the other "bad", are in conflict. In the former, prize-winning scientists are researching new nonpollutant sources of food and are in conflict with the multinational company Global Chemicals. Davros, creator of the Daleks, is totally amoral and unscrupulous, but his fellow Kaled scientists are willing to die in their attempt to stop his evil work.

Most of the images of women scientists (and there are more of them than in most contemporary science departments) are positive. Examples include Professor Laird in "Resurrection of the Daleks" (62), Professor Rachel Jensen and her assistant in "Remembrance of the Daleks" (63), and Professor Sarah Lasky in "Trial of a Time Lord—Terror of the Vervoids" (64). Although Professor Lasky was arrogant and mistaken, she gave her life to try to put things right. The motif of the scientist who has been "led astray", but then repents, recurs elsewhere—Dastari in "The Two Doctors" (65) and the mutants in "Mawdryn Undead" (66). Others are amoral and essentially mad, such as Taren Kapel, the roboticist of "Robots of Death" (67), Skagra in "Shada" (68), Eric Klieg, logician and world domination fanatic in "Tomb of the Cybermen" (69), or Stalman in "Inferno" (70). One of the most interesting is Sorensen in "Planet of Evil" (71). The mining operation on Zeta Minor is going badly wrong, and the minerals (which contain antimatter) effect a Dr. Jekyll/Mr. Hyde transformation on Sorensen. The Doctor convinces him of the error of his ways, and he eventually recovers. The Doctor's parting shot to him is a very powerful statement on responsible research; he reminds him that the right of a scientist to experiment requires that he take total responsibility for the outcome of the work.

In the original *Star Trek*, images of chemists are very few. The only one I could find was Marlena, of the alternative universe in "Mirror, Mirror" (72). She is certainly a positive and attractive character, but this derives from her physical rather than her intellectual charms. However, both McCoy and Spock are positively portrayed, and both do some excellent chemistry. McCoy finds the antidote to the tears of the women of Elaas in "Elaan of Troyius" (20) and cures the effect of an "area of space" with a derivative of thuragen, a Klingon nerve gas, in "The Tholian Web" (73). Spock finds the adrenaline antidote to the rapid aging process in "The Deadly Years" (74) with the aid of Dr. Janet Wallace, a biochemist. Captain Kirk's brother, George Samuel Kirk, is a biologist, as is his son David, and Sulu has a long-standing interest in botany.

In *Star Trek: The Next Generation*, the images of the crew are all very positive, and dramatic conflict usually arises from their interactions with the guest cast. Picard's scientific abilities are rarely stressed, except per-haps in "Chains of Command" (75), but in *Star Trek Generations* (76) he clearly takes great pride in the fact that one of his forebears won the Nobel Prize for chemistry. It is made clear that science in general, and chemistry in particular, forms an integral part of the curriculum at Starfleet Academy. In "The Game" (77), Picard reveals to Wesley that he failed organic chemistry as a result of his attraction for a young lady with the initials A. F. Geordi is an engineer, and Data has honors in exobiology and probability mechanics. Wesley, the wunderkind, is portrayed as being good at everything, but his accomplishments in math and physics are stressed. He does solve some interesting science problems in "Where No One Has Gone Before" (78), "Final Mission" (79), "Ménage à Troi" (80), and "Remember Me" (81), but most of these are physics or engineering, and the "Wesley saves the ship" episodes have never been particularly popular with the fans.

Worf's interests and expertise clearly do not lie in science, and although one might have expected that Troi, who has a degree in psychol-ogy, should have done some basic science, she seems lamentably ignorant. She fails to recognize a hydrogen atom in "Night Terrors" (42), and thinks that a cosmic string ["The Loss" (82)] is the same as a quantum filament ["Disaster" (83)]. Doctors Pulaski and Crusher do actually engage in some chemistry, despite their primary focus in medicine. For example, Beverly finds the antidote to the "water-carbon complex" in "The Naked Now" (2) and to the alkaloid poison in "Code of Honor" (84).

There are both positive and negative characterizations of scientists in the visiting casts, with the latter predominating. In "Home Soil" (27), the terraformers of Velara III were willing to destroy the inorganic life-form they discovered so that their work could progress. In "We'll Always Have Paris" (85), Dr. Paul Manheim is so excited about his discovery of a "crack" into another universe that he seems totally unconcerned that his entire scientific team have perished as a result of his experiments. After his own recovery he seems unchastened and is anxious only to return to his work.

The geneticists of Darwin Station in "Unnatural Selection" (34) also appear to be rather irresponsible; they have meddled with things they do not understand and seem very callous about the deaths of the crew of the U.S.S. *Lantree*. The Mariposans of "Up the Long Ladder" (86) are all clones; their ship had crashed, leaving only a few survivors, but because they were scientists and determined to survive, they had found a way to produce clones. In order to obtain new genetic material, they take cells from Geordi and Riker without their consent. The backward agrarian Bringloidi are clearly supposed to be the more attractive group, although they are portrayed in a patronizing and slightly racist manner. The contrast verges on caricature—chilly unemotional scientists who view conventional reproduction with disdain and disgust, and happy-go-lucky country folk, with normal healthy appetites.

Paul Stubbs, the physicist in "Evolution" (28) is also close to caricature. He must be based on someone's recollection of a particular professor; his appearance would attract no attention on our late 20th century physics department! He too cares more about his experiment than the safety of the ship, and he destroys many of the nanites, which have evolved into a sentient life-form.

In "The Hunted" (87), the Angosian scientists are portrayed unfavorably; they genetically and chemically altered their warriors to be more effective, but, once the war is over, have no interest in dealing with the problem they have created. Although this situation is depicted as a scientific problem, it may in fact be closer to a commentary on the feelings of Vietnam War veterans, who returned to a society in which many wanted to forget them, the reminder of their own uneasy consciences as to the desirability and conduct of the war.

Guinan is often shown acting as a conscience; she encourages Wesley to tell the captain what he has done in letting the nanites loose in "Evolution" (28). In their conversation it emerges that Wesley always gets an A for his courses. Guinan draws an analogy with another scientist of her acquaintance who did the same—a Dr. Frankenstein. She also supports Data's right to self-determination in "Measure of a Man" (88). She clearly has a "two cultures" point of view; in "Hollow Pursuits" (89) she says that engineers don't appreciate imagination. This remark is a little unfair; Geordi is interested in both art and music, and Barclay has a decidedly overactive imagination!

In "Silicon Avatar" (90) Dr. Kila Marr takes revenge for the death of her son at Omicron Theta by destroying the crystalline entity, just when the *Enterprise* had managed to communicate with it. Hannah Bates, high-powered theoretical physicist and blond babe in "The Masterpiece

Society" (91) and Nella Darren, stellar cartographer, talented musician and all-around hero in "Lessons" (92) are more sympathetic women scientists.

"Ethics" (93) is a thoughtful morality play and provides an opportunity to discuss the extent to which risk-taking in research is justified. Dr. Russell, who is visiting the *Enterprise*, wants to perform an experimental procedure on Worf after his back is broken in an accident. However, Dr. Crusher relieves her of duty after she kills a patient in testing a new theory. In the end she does treat Worf, and he recovers. Beverly rejoices in the outcome, but remains uneasy about the method. She considers that Russell takes too many short cuts to research success, at the expense of her patients' safety.

Dr. Farillon, in "The Quality of Life" (94) describes herself as a risk-taker, which she feels is uncharacteristic of most scientists, and she is also an obsessive workaholic. Although she is initially dismissive of the idea that the exocomps might be sentient, she eventually accepts that they are and will work with them on her project.

None of the scientists in "Suspicions" (95), except possibly the Ferengi, Dr. Reyga, is a sympathetic character, and one, Jo'bril, turns out to be a murderer. It's rare for the Ferengi to produce scientists (the only other one we've seen appears in "The Price" (96) in a sinister and underhand rôle), and the Klingons don't respect their scientists. In "Force of Nature" (97), the visiting scientists Drs. Rabal and Serova are attractive, if slightly obsessive, personalities. When Serova is not believed, she steals a shuttlecraft and sacrifices her life to prove her theory—that warp engines are causing damage to the Hekaris corridor. The local scientist in the medieval Barkonian culture in "Thine Own Self" (98) is initially hostile to Data, as he tries to correct her lesson that the elements are air, earth, fire, and water. However, on further acquaintance, she proves to be more open-minded and curious about his efforts to produce an anti-radiation drug.

The principal resident scientists on DS9 are Jadzia Dax, the science officer, and Julian Bashir, the medical officer. Both are attractive characters (at least once Bashir had overcome his arrogant and gauche behavior of the first season) and are portrayed as intelligent and innovative. However, neither gets involved in much chemistry. Miles O'Brien is more practical than either. In "Destiny" (99) he is told by the Cardassian scientist Delora that men have little aptitude for engineering so that on Cardassia women dominate the sciences. Surely, tongue-in-cheek political correctness? Science takes a bit of a battering on the station when Vedic Winn ["In the Hands of the Prophets" (*100*)] criticizes Keiko O'Brien for teaching the children about the science of the wormhole, which she regards as the celestial temple of her spiritual beliefs. She regards the lessons as blasphemous. Even First Officer Major Kira supports her, saying that teaching science without a moral context is teaching a philosophy. Should science be taught in a "value-free" mode? If not, whose values do we apply?

Lenara Khan is the female leader of the Trill science team in "Rejoined" (*101*); she is a positive character in general, though she loses her nerve toward the end. We discover that Curzon Dax, the previous host of the Dax symbiont, would be horrified by Jadzia's career in science.

The *Voyager* has a good complement of scientists. Captain Kathryn Janeway was a science officer before her promotion, and her commitment to research is obvious in "Resolutions (*102*) when she takes a protein analyzer and DNA sequencer to help her find a treatment for the viral infection that she and Chakotay have suffered. B'Elanna Torres, the former Maquis, is a highly creative and innovative engineer. She does some very neat work on the computer in "Dreadnought" (*103*) and outstanding robotics in "Prototype" (*104*). Harry Kim also seems to have good science credentials, and the holographic doctor has an inexhaustible supply of medical knowledge. In the first series, the episode "Jetrel" (*105*), broadcast close to the 50th anniversary of Hiroshima, is clearly intended as a metaphor for the involvement of scientists in the development of nuclear weapons. Dr. Ma'bor Jetrel was responsible for the metreon cascade, which wiped out more than a quarter of a million Talaxians. Although he designed the weapon, it was his government that decided to use it. He discusses the morality of this decision with Neelix, the Talaxian member of the *Voyager* crew, who saw the aftermath of the disaster. This follows a fairly predictable course; Jetrel believes that if he had not discovered the cascade, someone else would have done, and that for him all scientific knowledge is valuable. For Jetrel, the consequences were serious; his wife and children left him, and he contracted a fatal cancerlike illness from the radiation. Eventually Neelix manages to forgive him.

Does TV Reflect Contemporary Concerns and Issues?

The final part of this discussion deals with contemporary concerns and issues in science; do the series reflect these, and do they reflect them fairly? The early *Star Trek* original series episode, "Mudd's Women" (*106*), takes a light-hearted view of drug addiction. The Venus Drug, which Mudd supplies to his three lovelies, makes men more muscular and women more beautiful. The morally correct conclusion is reached; they do not need the drug, only their own self-confidence, but this story relates little to contemporary problems of drug abuse. It seems likely that it was never meant to do so; the original script of "The City on the Edge of Forever" (*107*) was intended to involve a drug addict (he rather than McCoy would go into the past and change history), but this idea was dropped as being too controversial.

In "Plato's Stepchildren" (*108*) (this episode was for long not shown on terrestrial TV in the United Kingdom; presumably the BBC, who persist in regarding *Star Trek* as a children's show, censored it because of the violence involved), the Platonians derive their telekinetic power from the kironide in their food. Again the morally correct conclusion is reached; Kirk and Spock take the drug to defeat the Platonians. The dwarf jester Alexander, who had a metabolic defect that meant he was not telekinetic, refuses the drug-induced power and opts instead to leave with the *Enterprise*. Again the drug issue is treated fairly lightly; this episode is chiefly remembered for the first interracial kiss to be shown on network TV in the United States.

A *Dr. Who* episode that treats drug addiction is "Nightmare of Eden" (*109*). The Doctor encounters smugglers of Vraxoin, the most dangerously addictive drug in the Galaxy. This episode treats the subject seriously and is one of the more thought-provoking stories. The effects of a drug trip, both the euphoric high and the symptoms of withdrawal, are depicted. One of the smugglers blames the addicts themselves for the trade; their weakness allows it to exist.

A similarly serious treatment is given in the TNG episode "Symbiosis" (*110*). The Brekkians produce felicium, which is a narcotic. The Ornarans believe it is a cure for the plague that affects them all, but in fact they are all addicts, a fact of which the Brekkians are very well aware.

This situation is quickly discovered by the *Enterprise* crew, but the responses of Picard and Dr. Crusher differ considerably. Picard upholds the Prime Directive; it is not their right to impose Federation values on others. Crusher, however, wants to help the Ornarans to overcome their addiction.

Wesley tries to understand why anyone should voluntarily become addicted to a chemical substance; his conversation with Data and Tasha is very clearly from the "Just Say No" era. Tasha explains the process of addiction, particularly in the context of the poverty and violence on her home planet. Wesley, not surprisingly, does not really understand. The interchange is perhaps a little cloying in delivering its moral message, but it is distinctly contemporary. The upshot is that the *Enterprise* does nothing; they do not help the Ornarans go cold turkey, and they don't repair their ship. It is old and decrepit and will obviously soon break down, so that there will be no more drug supplies. Crusher and Picard remain on opposite sides of the moral dilemma posed by the clash of the Prime Directive and medical ethics.

Neither *Star Trek: Deep Space Nine* nor *Star Trek: Voyager* have treated drug addiction very specifically; the best example is perhaps the "addiction" of Jem'Hadaar to ketresel white.

The aging process is a recurrent theme in many science fiction stories; various devices for prolonging life have their origins in ancient myths and reappear in medieval quest stories. In the TOS episode "Miri" (*111*), the *Enterprise* visits a planet where a life-prolongation project went horribly wrong. A virus killed the adult population very rapidly, and children contract it on entering puberty. However, until then, they age very slowly and are many hundreds of years old. The landing party contract the virus, but eventually find an antidote. In "The Omega Glory" (*26*), the *Enterprise* encounters a planet on which a natural immunizing agent allows individuals to live for a thousand years. The planet had, in the past, fought a bacteriological war (which is compared with one fought on Earth in the then-distant 1990s), and life prolongation is a result of the powerful protective antibodies developed by the population who survived. However, the planet has degenerated into near savagery; the price for long life has been high. How Flint became almost immortal in "Requiem for Methuselah" (*112*) (some 6000 years old) is never properly explained; he is, however, a sad and lonely figure. Rapid aging is seen in the story "The Deadly

Years" (*74*). A landing party of five beams down, and all but Chekov become infected with a virus that causes very rapid aging. The unknown Lt. Galway dies, because her metabolism caused her to age more rapidly. Fortunately a cure is found, and not only is the aging arrested, but actually reversed.

Dr. Who must be one of very few series that have survived a regular change of the actor playing the principal character. Regeneration is allowed up to 12 times for a Time Lord; the Doctor's archenemy, the Master, wants to extend this further, and succeeds in a further regeneration as a result of the power of the Keepership of Traken ["The Keeper of Traken" (*113*)]. There seem to be other exceptions to the rules; in "The Five Doctors" (*114*), the Master is offered another 12 regenerations by the High Council of the Time Lords in return for his help. In the same story, immortality is granted to Borusa in the tomb of Rassilon, but this is the immortality of eternal torment. The Sisterhood of Karn in "The Brain of Morbius" (*115*) use an elixir of youth, a chemical mixture produced in their volcano. The Doctor considers that this could easily be analyzed spectroscopically and its composition reproduced, but he stresses the stagnation that must inevitably result from a society of immortals. The Eternals in the story "Enlightenment" (*116*) don't seem to be having much fun out of their immortality. In "Mawdryn Undead" (*66*), a group of scientists come to an unhappy end when they try to acquire the Time Lords' power of regeneration, and Spectrox, a drug that can double life span, causes nothing but trouble in "The Caves of Androzani" (*117*).

As on most subjects, the TNG crew are thoroughly politically correct on the subject of the aging process. In "Too Short a Season" (*118*), Admiral Jameson takes an illegal rejuvenating drug so that he can complete his diplomatic mission and put right the mistakes he made in his youth. The outcome is disastrous, as his system cannot cope with the stress put upon it and he eventually dies. The episode ends with Riker and Picard lamenting the quest for youth.

In "The Neutral Zone" (*119*), Data finds three people from the 20th century who had been put into cryogenic stasis after death from causes that are readily treatable in the 24th century. They are revived and cured, and although they have problems adjusting to the idea of what has happened to them, they do seem to have achieved their original aim.

"Unnatural Selection" (34) tells a story involving rapid aging. The *Enterprise* comes across the U.S.S. *Lantree*, with all of its crew dead of extreme old age. From there they go on to a genetics research lab on Darwin Station, where they discover genetically engineered children whose immune systems are so active that they can alter the DNA of someone with whom they come into contact. This process causes transposition of two "DNA molecules", the genes that cause aging. As always, the final outcome is successful; Pulaski, who becomes infected because of her own foolhardiness, is "de-aged", using her DNA from a hair follicle and a clever trick with the transporter. The customary message is that we should be much more careful in meddling with things we do not fully understand. In "The Masterpiece Society" (91), Picard shows his disapproval of eugenics and genetic engineering. He feels that having a life that is essentially preordained restricts human potential and decreases originality and inventiveness.

Nuclear waste had not become so great an issue in the 1960s as it is today, and TOS does not deal with this problem. However, in the *Dr. Who* story "Genesis of the Daleks" (61), the planet Skaro has been heavily contaminated by the nuclear and chemical weapons used in the early years of the Kaled–Thaal wars. These weapons had caused genetic mutations in many of the population, and the Daleks were supposed to be the ultimate form into which the Kaled people would mutate. Also, in "Colony in Space" (120), the Doctor and Jo are dispatched to the barren planet Exarius 500 years in the future; the devastation has been caused by the use of a "doomsday weapon". In the TNG episode, "Final Mission" (79), a garbage scow with nuclear waste 300 years old is about to give a fatal dose of radiation to a planetary surface as it breaks up in the atmosphere. The *Enterprise* tows it out of the way and sends it into the sun just in time. In "Thine Own Self" (98), Data is sent to Barkon IV to recover radioactive material from a Federation deep-space probe, with near fatal results for the indigenous population. Again the message is clear—don't give hostages to the future.

Pollution in a more general sense has been a recurrent theme. The earliest mention comes in the *Dr. Who* episode "Planet of the Giants" (8), which is roughly contemporary with Rachel Carson's *Silent Spring (121)*. The Tardis travelers are accidentally shrunk in size and arrive in an English garden. Barbara gets a lethal dose of the insecticide DN-6, but recovers on being restored to normal size. The manufacturers of DN-6 are determined to release it even though they are aware that it will destroy all insect life. It is claimed that it has been made everlasting, although this seems relatively unlikely for an organophosphate. The story "The Green Death" (60) was specifically written to bring home the horrors of pollution. Global Chemicals, in a desire for productivity and efficiency, causes pollution that results in mutated green maggots with a fatal bite. In "The Mutants" (122), 30th century Earth is described as being so polluted that people live in caves or sky cities. The Terran Empire's mining activities have also polluted the colony planet Solos. Here daylight makes the polluted atmosphere toxic, presumably a reference to photochemical smog. "The Curse of Fenric" (123) presents a bleak picture of the future of humanity and Earth. The haemovores are the ultimate product of human evolution; after half a million years of industrial progress, Earth's surface is a chemical slime and the world is dying.

In several episodes of TOS there are semi-humorous asides about pollution in the mid-20th century. In each case ["A Piece of the Action" (124), "The City on the Edge of Forever" (107), and "Bread and Circuses" (125)], the crew have either gone back in time or are visiting a planet in a state of development comparable with that on 20th century Earth. In the TNG story "True Q" (126), the *Enterprise* is engaged in a relief effort to the ecologically devastated Tagran system. The Tagran atmosphere is badly polluted, and the crew is amazed that the Tagrans will go to great lengths to clean up the air with filters, rather than regulate the emissions that caused the pollution in the first place. Amanda, the "true Q" of the title, cleans up the pollution and restores the ecosystem. When Quark and his companions are transported to 20th-century Roswell in the *Deep Space Nine* episode "Little Green Men" (127), he expresses his horror that humans of that era poison themselves with tobacco, and their planet with fusion bombs.

Other environmental issues have also been discussed. In the *Dr. Who* story "Robot" (128), Professor Kettlewell's robot is made of a living metal that can be destroyed by a virus. This story was clearly provoked by concerns about waste metal, and there have certainly been proposals to engineer bacteria that could digest metals or plastic. Professor Kettlewell himself is motivated by a desire to stop mankind from polluting and destroying the planet. TNG has a reprise of this theme; in "A Matter of Honor" (*129*), a metal-eating bacterium makes a nasty hole in a Klingon vessel. In "Cost of Living" (*40*), a metal parasite attacks the alloy in the *Enterprise*, turning it into a gelatinous material. Any scientific pretensions that this episode might have had disappeared when Geordi said that the *elements* in the residue had been broken down into simpler *molecular* compounds!

Conservation is also a recurrent theme; most examples come from TNG. However, in the TOS story "The Man Trap" (*130*), the "salt vampire" is the last of its kind. Kirk reflects sadly on a comparison with the Earth buffalo, but kills the creature anyway.

In "The Devil in the Dark" (*24*), the Horta, also the last of her kind, is, however, allowed to survive. The *Star Trek* movies also take up this theme. *Star Trek IV* (*131*) is "Save the Whales", and *Star Trek VI* (*132*) is "Save the Klingons". "The Caretaker" (*133*), the pilot episode for *Star Trek: Voyager*, could be described as "Save the Ocampa"—the desert planet reminded me very much of *Dune* (*134*). The crew of the *Enterprise* in TNG are very environmentally conscious and distinctly politically correct. They are all said to be vegetarians, and many are quite disgusted by the idea of eating meat. Indeed most food is replicated rather than prepared, and what appears to be meat has had no connection with a living animal. Keiko is quite disgusted that Miles' mother cooked meat, and in "Where No One Has Gone Before" (*78*), Riker says piously that on Earth animals are no longer enslaved for food. However, in "Time Squared" (*135*), he cooks eggs that do not appear to have been replicated, and in "A Matter of Honor" (*129*) he has only modest difficulty with Klingon *gagh*, which is not only animal, but still alive. In "Sins of the Father" (*136*), Picard offers caviar to Kurn; this is apparently authentic, so perhaps not everyone is strictly vegan. The crew of *Voyager* have difficulty adjusting to "natural food", and replicator rations are much prized—but perhaps this is just the way Neelix cooks!

In "Galaxy's Child" (*137*), the crew is very upset that they have accidentally destroyed a space creature, and they make every effort to save its child. They also try hard to save Tin Man in the eponymous episode (*138*); they succeed, albeit in a manner they had not expected. In "New Ground" (*139*), we discover that white rhinos have become extinct on Earth, hunted because of the value of their horns. The *Enterprise* is transporting two Corvan gilvos to a new home because their natural habitat is threatened by industrial pollution of the rain forest.

The episode "When the Bough Breaks" (*140*) is interesting from an environmental point of view; the message once again is "Don't mess with things you don't understand." The ozone layer of the cloaked and invisible planet Aldea is failing, and the Aldeans have become sterile. They therefore steal especially talented children from the *Enterprise*. Gradually the reason for the Aldeans' problems becomes clear; they are suffering from chromosomal damage caused by radiation poisoning. This is compared with the problems faced by the Earth during the environmental deterioration of the 21st century, when the ozone layer failed and the UV level on the surface increased dramatically. Eventually the *Enterprise* team succeeds in penetrating and turning off the Aldeans' cloaking device and offers help in treating their condition, now that the children can be safely returned home. They successfully reseed the ozone layer, a clear reference to a number of suggestions that high-flying aircraft do the same on Earth. The Aldeans have survived but will never again be able to use their high-technology cloaking field.

The greenhouse effect is mentioned in the story "A Matter of Time" (*141*). The planet Penthara IV is hit by an asteroid, raising a dust cloud that causes an effect similar to the so-called "nuclear winter". The *Enterprise* finds carbon dioxide to create an artificial greenhouse effect to try to warm them up. The colonists say that they have been trying to avoid this effect for years, and indeed the attempt does go horribly wrong, as the drilling for CO_2 causes geological instability and a lot more dust.

The environmental messages are perhaps underlined a little too heavily, but there is some attempt at balance. In the *Dr. Who* story "Invasion of the Dinosaurs" (*142*), an environmental group wants to return to a so-called "golden age". However, they want to do this by turning back time and eliminating the whole development of the human race. The Doctor points out that the golden age they seek never really existed. He sees the real cause of pollution as greed. In the DS9 story "Paradise" (*143*), the simple nontechnological life is shown to be not exactly what it seemed. The architect of the trick, Alixus, is arrested, but most of the colonists decide to remain with the life they have built. One commentator (*144*) describes this episode as the nightmare of being shipwrecked with Margaret Thatcher.

Conclusions

The accuracy of the chemistry portrayed in these series may be improving, but not all the howlers have gone. Perhaps we are getting to the stage where these popular programs may have some use in the context of college and high school teaching.

However, the news is less good for chemistry than for other branches of science; few chemists are portrayed at all, and then with little sympathy. Scientists are becoming much more aware of, and concerned about, their public image. In 1994 Roslynn D. Haynes published an interesting study on this subject (*145*). Many of the scientists of literature, such as Moreau, Frankenstein, Caligari, Jekyll, and Strangelove, are amoral and often more than slightly mad. I am uncertain if the best-selling calendar for 1996, "Stud Muffins of Science", will do much to redress the balance. Chemistry receives much less attention than biology, physics, or engineering. The stories do reflect contemporary science to some extent, with a strong focus on areas with which the public finds it easy to identify, such as environmental concerns. Do the writers simply not know much about chemistry? Should we be educating them?

As science educators we cannot afford to ignore these and related TV programs. Their language and ideas have entered ours. In the *London Times* of September 28, 1992, an editorial begins by saying that a numismatic historian from Vulcan would conclude, by looking at British coinage since decimalization, that he was studying nation in decline. In the *Times Higher* of February 5, 1993, there is a review by Stuart Sutherland of "Irrationality; The Enemy Within" with the header "A species to baffle a Vulcan". The allusion is not explained in the text; it is expected that the reader will know. The terms have entered popular jargon. TV science fiction has the potential to turn the younger generation on to science; any series that depicts a six-year-old being scolded back to his calculus class, as this is something *everyone* must master ["When the Bough Breaks" (*140*)] has to be good for science education!

References

1. ST-TOS, "The Naked Time", John D. F. Black, Season 1, 1966.

2. ST-TNG, "The Naked Now", J. Michael Bingham from a story by John D. F. Black and J. Michael Bingham, Season 1, 1987.
3. Franks, F. *Polywater*; MIT Press: Boston, MA, 1981.
4. Vonnegut, Kurt, Jr. *Cat's Cradle*; Gollancz, London,1963.
5. Donovan, Paul. *The Sunday Times*, London, June 20, 1993.
6. "The Paradise of Death", Barry Letts, BBC radio, June 1993; *Dr. Who and the Paradise of Death*; Barry Letts, Virgin Publishing: London, 1994.;"The Ghosts of N-Space", Barry Letts, BBC radio, January 1996; *The Ghosts of N-Space*; Barry Letts, Virgin Publishing: London, 1995.
7. *Dr. Who*, "The Sensorites", Peter R. Newman, Season 1, 1964.
8. *Dr. Who*, "Planet of the Giants", Louis Marks, Season 2, 1964.
9. *Dr. Who*, "The Krotons", Robert Holmes, Season 6, 1968.
10. *Dr. Who*, "The Dominators", Norman Ashby, Season 6, 1968.
11. *Dr. Who*, "The Seeds of Death", Brian Hayles, Season 6, 1969.
12. *Dr. Who*, "The Claws of Axos", Bob Baker and Dave Martin, Season 8, 1971.
13. *Dr. Who*, "Planet of the Daleks", Terry Nation, Season 10, 1972–1973.
14. *Dr. Who*, "Revenge of the Cybermen", Gerry Davis, Season 12, 1975.
15. *Dr. Who*, "The Hand of Fear", Bob Baker and Dave Martin, Season 14, 1976.
16. *Dr. Who*, "The Stones of Blood", David Fisher, Season 16, 1978–1979.
17. *Dr. Who*, "Logopolis", Christopher H. Bidmead, Season 18, 1981.
18. *Dr. Who*, "Time and the Rani", Pip and Jane Baker, Season 24, 1987.
19. ST-TOS, "Where No Man Has Gone Before", Samuel A. Peeples, Season 1, 1966.
20. ST-TOS, "Elaan of Troyius", John Meredyth Lucas, Season 3, 1968.
21. ST-TOS, "The Galileo Seven", Oliver Crawford and S. Bar David from a story by Oliver Crawford, Season 1, 1966.
22. ST-TOS, "Space Seed", Gene L. Coon and Carey Wilber from a story by Carey Wilber, Season 1, 1967.
23. ST-TOS, "Arena", Gene L. Coon from a story by Fredric Brown, Season 1, 1967.
24. ST-TOS, "The Devil in the Dark", Gene L. Coon, Season 1, 1967.
25. ST-TOS, "Patterns of Force", John Meredyth Lucas, Season 2, 1968.
26. ST-TOS, "The Omega Glory", Gene Roddenberry, Season 2, 1968.
27. ST-TNG, "Home Soil", Robert Sabaroff, from a story by Karl Guers, Ralph Sanchez, and Robert Sabaroff, Season 1, 1988.
28. ST-TNG, "Evolution", Michael Piller from a story by Michael Piller and Michael Wagner, Season 3, 1989.
29. ST-TNG, "The Royale", Keith Mills (Pseudonym for Tracy Tormé), Season 2, 1989.
30. ST-TNG, "The Vengeance Factor", Sam Rolfe, Season 3, 1989.
31. ST-TNG, "Suddenly Human", John Whelpley and Jeri Taylor from a story by Ralph Phillips, Season 4, 1990.

32. ST-TNG, "Legacy", Joe Menosky, Season 4, 1990.

33. ST-TNG, "Data's Day", Harold Apter and Ronald D. Moore from a story by Harold Apter, Season 4, 1991.

34. ST-TNG, "Unnatural Selection", John Mason and Mike Gray, Season 2, 1989.

35. ST-TNG, "Transfigurations", René Echevarria, Season 3, 1990.

36. ST-TNG, "The Drumhead", Jeri Taylor, Season 4, 1991.

37. ST-TNG, "The Chase", Joe Menosky from a story by Ronald D. Moore, Season 6, 1993.

38. ST-TNG, "Sub Rosa", Brannon Braga from a story by Jeri Taylor, Season 7, 1993.

39. ST-TNG, "Genesis", Brannon Braga, Season 7, 1994.

40. ST-TNG, "Cost of Living", Peter Allan Fields, Season 5, 1992.

41. ST-TNG, "Schisms", Brannon Braga from a story by Jean Louise Matthias and Ron Wilkerson, Season 6, 1992.

42. ST-TNG, "Night Terrors", Pamela Douglas and Jeri Taylor from a story by Shari Goodhartz, Season 4, 1991.

43. ST-TNG, "Firstborn", René Echevarria from a story by Mark Kalbfeld, Season 7, 1994.

44. ST-TNG, "Time's Arrow", Joe Menosky and Michael Piller from a story by Joe Menosky, Season 5, 1992.

45. ST-DS9, "A Man Alone", Michael Piller from a story by Gerald Sandford and Michael Piller, Season 1, 1993.

46. ST-DS9, "The Alternate", Bill Dial from a story by Jim Trombetta and Bill Dial, Season 2, 1994.

47. ST-DS9, "Broken Link", Robert Hewitt Wolfe and Ira Stephen Behr from a story by George A. Brozak, Season 4, 1996.

48. ST-DS9, "The Abandoned", D. Thomas Maio and Steve Warnek, Season 3, 1994.

49. Crichton, Michael, *Jurassic Park*; Random House: New York, 1991.

50. ST-DS9, "Hippocratic Oath", Lisa Klink from a story by Nicholas Corea and Lisa Klink, Season 4, 1995.

51. ST-DS9, "To the Death", Ira Stephen Behr and Robert Hewitt Wolfe, Season 4, 1996.

52. ST-V, "Emanations", Brannon Braga, Season 1, 1995.

53. ST-V, "Learning Curve", Ronald Wilkerson and Jean Louise Matthias, Season 1, 1995.

54. ST-V, "The 37s", Jeri Taylor and Brannon Braga, Season 2, 1995.

55. ST-V, "Cold Fire", Brannon Braga from a story by Anthony Williams, Season 2, 1995.

56. ST-V, "Basics—Part 1", Michael Piller, Season 2, 1996.

57. ST-V, "Threshold", Brannon Braga from a story by Michael de Luca, Season 2, 1996.

58. ST-V, "Tuvix", Kenneth Biller from a story by Andrew Shepherd and Mark Gaberman, Season 2, 1996.

59. Dr. Who, "The Three Doctors", Bob Baker and Dave Martin, Season 10, 1972–1973.

60. Dr. Who, "The Green Death", Robert Sloman, Season 10, 1972–1973.

61. Dr. Who, "Genesis of the Daleks", Terry Nation, Season 12, 1975

62. Dr. Who, "Resurrection of the Daleks", Eric Saward, Season 21, 1984.

63. Dr. Who, "Remembrance of the Daleks", Ben Aaronovitch, Season 25, 1988.

64. Dr. Who, "Trial of a Time Lord—Terror of the Vervoids", Pip and Jane Baker, Season 23, 1986.

65. Dr. Who, "The Two Doctors", Robert Holmes, Season 22, 1985.

66. Dr. Who, "Mawdryn Undead", Peter Grimwade, Season 20, 1983.

67. Dr. Who, "Robots of Death", Chris Boucher, Season 14, 1976–1977.

68. Dr. Who, "Shada", Douglas Adams, Season 17 (not transmitted but part is available on video), 1979–1980.

69. Dr. Who, "Tomb of the Cybermen", Kit Pedler and Gerry Davis, Season 5, 1967–1968.

70. Dr. Who, "Inferno", Don Houghton, Season 7, 1970.

71. Dr. Who, "Planet of Evil", Robert Banks Stewart, Season 13, 1975–1976.

72. ST-TOS, "Mirror, Mirror", Jerome Bixby, Season 2, 1967.

73. ST-TOS, "The Tholian Web", Judy Burns and Chet Richards, Season 3, 1968.

74. ST-TOS, "The Deadly Years", David P. Harmon, Season 2, 1967.

75. ST-TNG, "Chains of Command", Ronald D. Moore from a story by Frank Abatemarco, Season 6, 1992.

76. Star Trek Generations, Rick Berman, Ronald D. Moore, and Brannon Braga, Paramount Pictures, 1994.

77. ST-TNG, "The Game", Brannon Braga from a story by Susan Sackett, Fred Bronson, and Brannon Braga, Season 5, 1991.

78. ST-TNG, "Where No One Has Gone Before", Diane Duane, Michael Reaves, and Maurice Hurley, Season 1, 1987.

79. ST-TNG, "Final Mission", Kacey Arnold-Ince and Jeri Taylor, from a story by Kacey Arnold-Ince, Season 4, 1990.

80. ST-TNG, "Ménage à Troi", Fred Bronson and Susan Sackett, Season 3, 1990.

81. ST-TNG, "Remember Me", Lee Sheldon, Season 4, 1990.

82. ST-TNG, "The Loss", Hilary J. Bader, Alan J. Adler, and Vanessa Greene from a story by Hilary J. Bader, Season 4, 1990.

83. ST-TNG, "Disaster", Ronald D. Moore from a story by Ron Jarvis and Philip A. Scorza, Season 5, 1991.

84. ST-TNG, "Code of Honor", Kathryn Powers and Michael Baron, Season 1, 1987.

85. ST-TNG, "We'll Always Have Paris", Deborah Dean Davis and Hannah Louise Shearer, Season 1, 1988.

86. ST-TNG, "Up the Long Ladder", Melinda M. Snodgrass, Season 2, 1989.
87. ST-TNG, "The Hunted", Robert Bernheim, Season 3, 1990.
88. ST-TNG, "Measure of a Man", Melinda M. Snodgrass, Season 2, 1989.
89. ST-TNG, "Hollow Pursuits", Sally Caves, Season 3, 1990.
90. ST-TNG, "Silicon Avatar", Jeri Taylor from a story by Lawrence V. Conley, Season 5, 1991.
91. ST-TNG, "The Masterpiece Society", Adam Belanoff and Michael Piller from a story by James Kahn and Adam Belanoff, Season 5, 1992.
92. ST-TNG, "Lessons", Ronald Wilkerson and Jean Louise Matthias, Season 6, 1993.
93. ST-TNG, "Ethics", Ronald D. Moore from a story by Sara Chamo and Stuart Chamo, Season 5, 1992.
94. ST-TNG, "The Quality of Life", Naren Shankar, Season 6, 1992.
95. ST-TNG, "Suspicions", Joe Menosky and Naren Shankar, Season 6, 1993.
96. ST-TNG, "The Price", Hannah Louise Shearer, Season 3, 1989.
97. ST-TNG, "Force of Nature", Naren Shankar, Season 7, 1993.
98. ST-TNG, "Thine Own Self", Ronald D. Moore from a story by Christopher Hatton, Season 7, 1994.
99. ST-DS9, "Destiny", David S. Cohen and Martin A. Winder, Season 3, 1995.
100. ST-DS9, "In the Hands of the Prophets", Robert Hewitt Wolfe, Season 1, 1993.
101. ST-DS9, "Rejoined", Ronald D. Moore and René Echevarria from a story by René Echevarria, Season 4, 1995.
102. ST-V, "Resolutions", Jeri Taylor, Season 2, 1996.
103. ST-V, "Dreadnought", Gary Holland, Season 2, 1996.
104. ST-V, "Prototype", Nicholas Corea, Season 2, 1996.
105. ST-V, "Jetrel", Jack Klein, Karen Klein, and Kenneth Biller from a story by James Thornton and Scott Nimerfro, Season 1, 1995.
106. ST-TOS, "Mudd's Women", Stephen Kandel from a story by Gene Roddenberry, Season 1, 1966.
107. ST-TOS, "The City on the Edge of Forever", Harlan Ellison, Season 1, 1967.
108. ST-TOS, "Plato's Stepchildren", Meyer Dolinsky, Season 3, 1969.
109. *Dr. Who*, "Nightmare of Eden", Bob Baker, Season 17, 1979–1980.
110. ST-TNG, "Symbiosis", Robert Lewin, Richard Manning, and Hans Beimler from a story by Robert Lewin, Season 1, 1988.
111. ST-TOS, "Miri", Adrian Spies, Season 1, 1966.
112. ST-TOS, "Requiem for Methuselah", Jerome Bixby, Season 3, 1969.
113. *Dr. Who*, "The Keeper of Traken", Johnny Byrne, Season 18, 1980–1981.
114. *Dr. Who*, "The Five Doctors", Terrance Dicks, Season 20, 1983.
115. *Dr. Who*, "The Brain of Morbius", Robin Bland, Season 13, 1975–1976.
116. *Dr. Who*, "Enlightenment", Barbara Clegg, Season 20, 1983.
117. *Dr. Who*, "The Caves of Androzani", Robert Holmes, Season 21, 1984.

118. ST-TNG, "Too Short a Season", Michael Michaelian and D. C. Fontana from a story by Michael Michaelian, Season 1, 1988.
119. ST-TNG, "The Neutral Zone", Maurice Hurley from a story by Deborah McIntyre and Mona Glee, Season 1, 1988.
120. *Dr. Who*, "Colony in Space", Malcolm Hulke, Season 8, 1971.
121. Carson, Rachel. *Silent Spring*, Houghton Mifflin: New York, 1960.
122. *Dr. Who*, "The Mutants", Bob Baker and Dave Martin, Season 9, 1972.
123. *Dr. Who*, "The Curse of Fenric", Ian Briggs, Season 26, 1989.
124. ST-TOS, "A Piece of the Action", David P. Harmon and Gene L. Coon from a story by David P. Harmon, Season 2, 1968.
125. ST-TOS, "Bread and Circuses", Gene Roddenberry and Gene L. Coon, Season 2, 1968.
126. ST-TNG, "True Q", René Echevarria, Season 6, 1992.
127. ST-DS9, "Little Green Men", Ira Stephen Behr and Robert Hewitt Wolfe from a story by Tony Marberry and Jack Trevino, Season 4, 1995.
128. *Dr. Who*, "Robot", Terrance Dicks, Season 12, 1974–1975.
129. ST-TNG, "A Matter of Honor", Burton Armus from a story by Wanda M. Haight, Gregory Amos, and Burton Armus, Season 2, 1989.
130. ST-TOS, "The Man Trap", George Clayton Johnson, Season 1, 1966.
131. *Star Trek IV: The Voyage Home*, screenplay by Steve Meerson, Peter Krikes, Harve Bennet, and Nicholas Meyer, story by Leonard Nimoy and Harve Bennet, Paramount Pictures, 1987.
132. *Star Trek VI: The Undiscovered Country*, screenplay by Nicholas Meyer and Denny Martin Finn, story by Leonard Nimoy, Lawrence Honner, and Mark Rosenthal, Paramount Pictures, 1991.
133. ST-V, "The Caretaker", Michael Piller and Jeri Taylor from a story by Michael Piller, Jeri Taylor, and Rick Berman, Season 1, 1995.
134. Herbert, Frank. *Dune*, Gollancz: 1967.
135. ST-TNG, "Time Squared", Maurice Hurley from a story by Kurt Michael Bensmiller, Season 2, 1989.
136. ST-TNG, "Sins of the Father", Ronald D. Moore and W. Reed Morgan, based on a teleplay by Drew Deighan, Season 3, 1990.
137. ST-TNG, "Galaxy's Child", Maurice Hurley from a story by Thomas Kartozian, Season 4, 1991.
138. ST-TNG, "Tin Man", Dennis Putman Bailey and David Bischoff, Season 3, 1990.
130. ST-TNG, "New Ground", Grant Rosenberg from a story by Sara Chamo and Stuart Chamo, Season 5, 1992.
140. ST-TNG, "When the Bough Breaks", Hannah Louise Shearer, Season 1, 1988.
141. ST-TNG, "A Matter of Time", Rick Berman, Season 5, 1991.
142. *Dr. Who*, "Invasion of the Dinosaurs", Malcolm Hulke, Season 11, 1973–1974.

143. ST-DS9, "Paradise", Jeff King, Richard Manning, and Hans Beimler from a story by Jim Trombetta and James Crocker, Season 2, 1994.

144. Cornell, P.; Day, M.; Topping, K., *The New Trek Programme Guide*; Virgin Books: London, 1995.

145. Haynes, Roslynn D. *From Faust to Dr. Strangelove: Representations of the Scientist in Western Literature*; The Johns Hopkins University Press: Baltimore, MD, 1994.

Chemists at Play

12

The Endochronic Properties of Resublimated Thiotimoline

Isaac Asimov

★ *An article on a remarkable chemical substance. This one seems to have escaped from the J.A.C.S.—probably by request. (John W. Campbell, Editor)*

The correlation of the structure of organic molecules with their various properties, physical and chemical, has in recent years afforded much insight into the mechanism of organic reactions, notably in the theories of resonance and mesomerism as developed in the last decade. The solubilities of organic compounds in various solvents has become of particular interest in this connection through the recent discovery of the endochronic nature of thiotimoline.[1]

It has been long known that the solubility of organic compounds in polar solvents such as water is enhanced by the presence upon the hydrocarbon nucleus of hydrophilic —i.e. water-loving—groups, such as the hydroxy (-OH), amino (-NH$_2$), or sulfonic acid (SO$_3$H) groups. Where the physical characteristics of two given compounds—particularly the degree of subdivision of the material—are equal, then the time of solution—expressed in seconds per gram of material per milliliter of solvent—decreases with the number of hydrophilic groups present. Catechol, for instance, with two hydroxy groups on the benzene nucleus dissolves considerably more quickly than does phenol with only one hydroxy group on the nucleus. Feinschreiber and Hravlek[2] in their studies on the problem have contended that with increasing hydrophilism, the time of solution approaches zero. That this analysis is not entirely correct was shown when it was discovered that the compound thiotimoline will dissolve in water —in the proportions of 1 gm./ml.— in *minus* 1.12 seconds. That is, it will dissolve *before* the water is added.

Previous communications from these laboratories indicated thiotimoline to contain at least fourteen hydroxy groups, two amino groups and one sulfonic acid group.[3] The presence of a nitro group ($-NO_2$) in addition has not yet been confirmed and no evidence as yet exists as to the nature of the hydrocarbon nucleus, though an at least partly aromatic structure seems certain.

The Endochronometer—First attempts to measure the time of solution of thiotimoline quantitatively met with considerable difficulty because of the very negative nature of the value. The fact that the chemical dissolved prior to the addition of the water made the attempt natural to withdraw the water after solution and before addition. This, fortunately for the law of Conservation of Mass-Energy, never succeeded since solution never took place unless the water was eventually added. The question is, of course, instantly raised as to how the thiotimoline can "know" in advance whether the water will ultimately be added or not. Though this is not properly within our province as physical chemists, much recent material has been pub-

lished within the last year upon the psychological and philosophical problems thereby posed.[4,5]

Nevertheless, the chemical difficulties involved rest in the fact that the time of solution varies enormously with the exact mental state of the experimenter: A period of even slight hesitation in adding the water reduces the negative time of solution, not infrequently wiping it out below the limits of detection. To avoid this, a mechanical device has been constructed, the essential design of which has already been reported in a previous communication.[6] This device, termed the endochronometer, consists of a cell 2 cubic centimeters in size into which a desired weight of thiotimoline is placed, making certain that a small hollow extension at the bottom of the solution cell—1 millimeter in internal diameter—is filled. To the cell, is attached an automatic pressure micro-pipette containing a specific volume of the solvent concerned. Five seconds after the circuit is closed, this solvent is automatically delivered into the cell containing the thiotimoline. During the time of action, a ray of light is focused upon the small cell-extension described above, and at the instant of solution, the transmission of this light will no longer be impeded by the presence of solid thiotimoline. Both the instant of solution—at which time the transmission of light is recorded by a photoelectric device—and the instant of solvent addition can be determined with an accuracy of better than 0.01%. If the first value is subtracted from the second, the time of solution (T) can be determined.

The entire process is conducted in a thermostat maintained at 25.00° C.—to an accuracy of 0.01° C.

Thiotimoline Purity—The extreme sensitivity of this method highlights the deviations resulting from trifling impurities present in thiotimoline. (Since no method of laboratory synthesis of the substance has been devised, it may be practically obtained only through tedious isolation from its natural source, the bark of the shrub *Rosacea Karlsbadensis rufo.*[7]) Great efforts were therefore made to purify the material through repeated recrystallizations from conductivity water—twice re-distilled in an all-tin apparatus—and through final sublimations. A comparison of the solution times (T) at various stages of the purification process is shown in Table I.

It is obvious from Table I that for truly quantitative significance, thiotimoline purified as described

must be used. After the second resublimation, for instance, the error involved in an even dozen determinations is less than 0.7% with the extreme values being —1.119 seconds and —1.126 seconds.

In all experiments described subsequently in this study, thiotimoline so purified has been used.

Time of Solution and Volume of Solvent—As would seem reasonable, experiments have shown that increasing the volume of solvent enables the thiotimoline to dissolve more quickly—i.e., with an increasingly negative time of solution. From Figure 1, however, we can see that this increase in endochronic properties levels off rapidly after a volume of solvent of approximately 1.25 ml. This interesting plateau effect has appeared with varying volume of solvent for all varieties of solvents used in these laboratories, just as in all cases the time of solution approaches zero with decreasing volume of solvent.

TABLE I

Purification Stage	Average "T" (12 observations)	"T" extremes	% error
As Isolated	—0.72	—0.25; —1.01	34.1
First recrystallization	—0.95	—0.84; —1.09	9.8
Second recrystallization	—1.05	—0.99; —1.10	4.0
Third recrystallization	—1.11	—1.08; —1.13	1.8
Fourth recrystallization	—1.12	—1.10; —1.13	1.7
First resublimation	—1.12	—1.11; —1.13	0.9
Second resublimation	—1.122	—1.12; —1.13	0.7

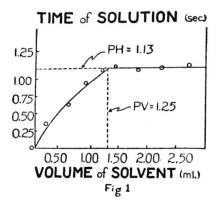

Fig 1

with decreasing concentration of sodium chloride, approaching the PV for water as the NaCl concentration approaches zero. It is, therefore, obvious that a sodium chloride solution of unknown concentration can be quite accurately characterized by the determination of its PV, where other salts are absent.

This usefulness of PV extends to other ions as well. Figure 3 gives the endochronic curves for 0.001 molar solutions of sodium chloride,

Time of Solution and Concentration of a Given Ion—In Figure 2, the results are given of the effect of the time of solution (T) of varying the volume of solvent, where the solvent consists of varying concentrations of sodium chloride solution. It can be seen that although in each case, the volume at which this plateau is reached differs markedly with the concentration, the heights of the plateau are constant (i.e. —1.13). The volume at which it is reached, hereinafter termed the Plateau Volume (PV), decreases

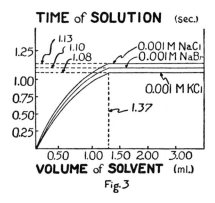

Fig.3

sodium bromide, and potassium chloride. Here, the PV in each case is equal within the limits of experimental error—since the concentrations in each case are equal— but the Plateau Heights (PH) *are* different.

A tentative conclusion that might be reached from this experimental data is that the PH is characteristic of the nature of the ions present in solution whereas the PV is characteristic of the concentration of these ions. Table II gives the values of Plateau Height and Plateau Vol-.

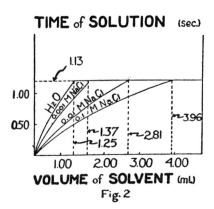

Fig.2

ume for a wide variety of salts in equal concentrations, when present alone.

The most interesting variation to be noted in Table II is that of the PV with the valence type of the salt present. In the case of salts containing pairs of singly-charged ions —i.e. sodium chloride, potassium chloride, and sodium bromide—the PV is constant for all. This holds also for those salts containing one singly charged ion and one doubly charged ion—i.e. sodium sulphate, calcium chloride, and magnesium chloride—where the PV, though equal among the three varies markedly from those of the first set. The PV is, therefore, apparently a function of the ionic strength of the solution.

This effect also exists in connection with the Plateau Height, though less regularly. In the case of singly charged ions, such as in the first three salts listed in Table II, the PH is fairly close to that of water itself. It falls considerably where doubly charged ions, such as sul-

phate or calcium are present. And when the triply charged phosphate ion or ferric ion is present, the value sinks to merely a quarter of its value in water.

Time of Solution and Mixtures of Ions—Experiments currently in progress in these laboratories are concerned with the extremely important question of the variation of these endochronic properties of thiotimoline in the presence of mixtures of ions. The state of our data at present does not warrant very general conclusions, but even our preliminary work gives hope of the further development of the endochronic methods of analysis. Thus, in Figure 4, we have the endochronic curve where a mixture of 0.001M Sodium Chloride and 0.001 Ferric Chloride solutions is the solvent. Here, two sharp changes in slope can be seen: the first at a solution time of —0.29, and the second at —1.13, these being the PH's characteristic of Ferric Chloride and Sodium Chloride respectively—see

TABLE II

Solvent (Salt solutions in 0.001 M concentration)	Plateau Height (PH) seconds	Plateau Volume (PV) milliliters
Water	—1.13	1.25
Sodium Chloride solution	—1.13	1.37
Sodium Bromide solution	—1.10	1.37
Potassium Chloride solution	—1.08	1.37
Sodium Sulphate solution	—0.72	1.59
Calcium Chloride solution	—0.96	1.58
Magnesium Chloride solution	—0.85	1.59
Calcium Sulphate solution	—0.61	1.72
Sodium Phosphate solution	—0.32	1.97
Ferric Chloride solution	—0.29	1.99

Table II. The PH for a given salt would thus appear not to be affected by the presence of other salts.

This is definitely not the case, however, for the PV, and it is to a quantitative elucidation of the variation of PV with impurities in the solvent that our major efforts are now directed.

TIME of SOLUTION (sec.)

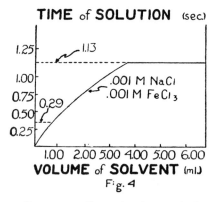

VOLUME of SOLVENT (ml)

Fig. 4

Summary—Investigations of the endochronic qualities of thiotimoline have shown that:

a—Careful purification of the material is necessary for obtaining quantitative results.

b—Increasing the volume of solvent results in increasing the negative time of solution to a constant value known as the Plateau Height (PH), at a volume of solvent known as the Plateau Volume (PV).

c—The value of the PH is characteristic of the nature of the ions present in the solvent, varying with the ionic strength of the solution and not varying with the addition of other ions.

d—The value of the PV is characteristic of the concentration of the ions present in the solvent, being constant for different ions in solution of equal ionic strength, but varying markedly with the admixtures of second varieties of ions.

As a result of all this, it is suggested that endochronic methods offer a means of rapid—2 minutes or less—and accurate—within 0.1% at least—analysis of inorganic, water-soluble materials.

Bibliography:

P. Krum and L. Eshkin, *Journal of Chemical Solubilities, 27,* 109-114 (1944). "Concerning the Anomalous Solubility of Thiotimoline."

E. J. Feinschreiber and Y. Hravlek. *Journal of Chemical Solubilities, 22,* 57-68 (1939), "Solubility Speeds and Hydrophilic Groupings."

P. Krum, L. Eshkin, and O. Nile. *Annals of Synthetic Chemistry, 115,* 1122-1145; 1208-1215 (1945), "Structure of Thiotimoline, Parts I & II."

G. H. Freudler, *Journal of Psychochemistry, 2,* 476-488 (1945), "Initiative and Determination: Are They Influenced by Diet?—As tested by Thiotimoline solubility Experiments."

E. Harley-Short, *Philosophical Proceedings & Reviews, 15,* 125-197 (1946). "Determinism and Free-Will. The Application of Thiotimoline Solubility to Marxian Dialectic."

P. Krum, "*Journal of Chemical Solubilities, 29,* 818-819 (1946), "A Device for the Quantitative Measurement of Thiotimoline Solubility Speed."

A. Roundin, B. Lev, and Y. J. Prutt, *Proceedings of the Society of Plant Chemistry, 80,* 11-18 (1930), "Natural Products isolated from shrubs of the genus *Rosacea.*"

Tiotimolin kak Ispitatel Markscnskoy dilektiki B. Kreschiatika, *Journal Naouki i Sovetskoy Ticorii* Vol. 11, No. 3.

Philossophia Neopredelennosti i Tiotimolin, Molvinski Pogost i Z. Brikalo, *Mir i Kultura* Vol. 2, No. 31.

THE END.

★13★

Pâté de Foie Gras

Isaac Asimov
Illustrated by Freas

★*Your kind attention is specially invited to this Special Feature! Let's see you crack this item! (John W. Campbell, Editor)*

I couldn't tell you my real name if I wanted to and, under the circumstances, I don't want to.

I'm not much of a writer myself, unless you count the kind of stuff that passes muster in a scientific paper, so I'm having Isaac Asimov write this up for me.

I've picked him for several reasons. First, he's a biochemist, so he understands what I tell him; some of it, anyway. Secondly, he can write; or at least he has published considerable fiction, which may not, of course, be the same thing.

But most important of all, he can get what he writes published in sci-ence-fiction magazines and he has written two articles on thiotimoline, and that is exactly what I need for reasons that will become clear as we proceed.

I was not the first person to have the honor of meeting The Goose. That belongs to a Texas cotton-farmer named Ian Angus Mac-Gregor, who owned it before it became government property. (The names, places and dates I use are deliberately synthetic. None of you will be able to trace anything through them. Don't bother trying.)

MacGregor apparently kept geese

about the place because they ate weeds, but not cotton. In this way, he had automatic weeders that were self-fueling and, in addition, produced eggs, down, and, at judicious intervals, roast goose.

By summer of 1955, he had sent an even dozen of letters to the Department of Agriculture requesting information on the hatching of goose eggs. The department sent him all the booklets on hand that were anywhere near the subject, but his letters simply got more impassioned and freer in their references to his "friend," the local Congressman.

My connection with this is that I am in the employ of the Department of Agriculture. I have considerable training in agricultural chemistry, plus a smattering of vertebrate physiology. (This won't help you. If you think you can pin my identity out of this, you are mistaken.)

Since I was attending a convention at San Antonio in July of 1955, my boss asked me to stop off at Mac-Gregor's place and see what I could do to help him. We're servants of the public and besides we had finally received a letter from MacGregor's congressman.

On July 17, 1955, I met The Goose.

I met MacGregor first. He was in his fifties, a tall man with a lined face full of suspicion. I went over all the information he had been given, explained about incubators, the values of trace minerals in the diet, plus some late information on

Vitamin E, the cobalamins and the use of antibiotic additives.

He shook his head. He had tried it all and still the eggs wouldn't hatch.

What could I do? I'm a Civil Service employee and not the archangel, Gabriel. I'd told him all I could and if the eggs still wouldn't hatch, they wouldn't and that was that. I asked politely if I might see his geese, just so no one could say afterward I hadn't done all I possibly could.

He said, "It's not geese, mister; it's one goose."

I said, "May I see the one goose?"

"Rather not."

"Well, then, I can't help you any further. If it's only one goose, then there's just something wrong with it. Why worry about one goose? Eat it."

I got up and reached for my hat.

He said, "Wait!" and I stood there while his lips tightened and his eyes wrinkled and he had a quiet fight with himself.

He said, "If I show you something, will you swear to keep it secret?"

He didn't seem like the type of man to rely on another's vow of secrecy, but it was as though he had reached such a pit of desperation that he had no other way out.

I said, "If it isn't anything criminal—"

"Nothing like that," he snapped.

And then I went out with him to a pen near the house, surrounded by barbed wire, with a locked gate to it, and holding one goose—The Goose. "That's The Goose," he said. The way he said it, I could hear the capitals.

I stared at it. It looked like any other goose, Heaven help me, fat, self-satisfied and short-tempered. I said, "Hm-m-m" in my best professional manner.

MacGregor said, "And here's one of its eggs. It's been in the incubator. Nothing happens." He produced it from a capacious overall pocket. There was a queer strain about his manner of holding it.

I frowned. There was something wrong with the egg. It was smaller and more spherical than normal.

MacGregor said, "Take it."

I reached out and took it. Or tried to. I gave it the amount of heft an egg like that ought to deserve and it just sat where it was. I had to try harder and then up it came.

Now I knew what was queer about the way MacGregor held it. It weighed nearly two pounds. (To be exact, when we weighed it later, we found its mass to be 852.6 grams.)

I stared at it as it lay there, pressing down the palm of my hand, and MacGregor grinned sourly. "Drop it," he said.

I just looked at him, so he took it out of my hand and dropped it himself.

It hit soggy. It didn't smash. There was no spray of white and yolk. It just lay where it fell with the bottom caved in.

I picked it up again. The white eggshell had shattered where the egg had struck. Pieces of it had flaked away and what shone through was a dull yellow in color.

My hands trembled. It was all I could do to make my fingers work, but I got some of the rest of the shell flaked away, and stared at the yellow.

I didn't have to run any analyses. My heart told me.

I was face to face with The Goose!

The Goose That Laid The Golden Eggs!

You don't believe me. I'm sure of that. You've got this tabbed as another thiotimoline article.

Good! I'm *counting* on your thinking that. I'll explain later.

Meanwhile, my first problem was to get MacGregor to give up that golden egg. I was almost hysterical about it. I was almost already to clobber him and make off with the egg by force if I had to.

I said, "I'll give you a receipt. I'll guarantee you payment. I'll do anything in reason. Look, Mr. MacGregor, they're no good to you anyway. You can't cash the gold unless you can explain how it came into your possession. Holding gold is illegal. And how do you expect to explain? If the government—"

"I don't want the government butting in," he said, stubbornly.

But I was twice as stubborn. I followed him about. I pleaded. I yelled. I threatened. It took me hours. Literally. In the end, I signed a receipt and he dogged me out to my car and stood in the road as I drove away, following me with his eyes.

He never saw that egg again. Of course, he was compensated for the value of the gold—$656.47 after taxes had been subtracted—but that was a bargain for the government.

When one considers the potential value of that egg—

The *potential* value! That's the irony of it. That's the reason for this article.

The head of my section at the Department of Agriculture is Louis P. Bronstein. (Don't bother looking him up. The "P." stands for Pittfield if you want more misdirection.)

He and I are on good terms and I felt I could explain things without being placed under immediate observation. Even so, I took no chances. I had the egg with me and when I got to the tricky part, I just laid it on the desk between us.

Finally, he touched it with his finger as though it were hot.

I said, "Pick it up."

It took him a long time, but he did, and I watched him take two tries at it as I had.

I said, "It's a yellow metal and it could be brass only it isn't because it's inert to concentrated nitric acid. I've tried that already. There's only a shell of gold because it can be bent with moderate pressure. Besides, if it were solid gold, the egg would weigh over ten pounds."

Bronstein said, "It's some sort of hoax. It *must* be."

"A hoax that uses real gold? Re-member, when I first saw this thing, it was covered completely with authentic unbroken eggshell. It's been easy to check a piece of the eggshell. Calcium carbonate. That's a hard thing to gimmick. And if we look inside the egg—I didn't want to do that on my own, chief—and find real egg, then we've got it, because that would be impossible to gimmick. Surely, this is worth an official project."

"How can I approach the Secretary with—" He stared at the egg.

But he did in the end. He made phone calls and sweated out most of a day. One or two of the department brass came to look at the egg.

Project Goose was started. That was July 20, 1955.

I was the responsible investigator to begin with and remained in titular charge throughout, though matters quickly got beyond me.

We began with the one egg. Its average radius was 35 millimeters (major axis, 72 millimeters; minor axis, 68 millimeters). The gold shell was 2.45 millimeters in thickness. Studying other eggs later on, we found this value to be rather high. The average thickness turned out to be 2.1 millimeters.

Inside *was* egg. It looked like egg and it smelled like egg.

Aliquots were analyzed and the organic constituents were reasonably normal. The white was 9.7 per cent albumin. The yolk had the normal complement of vitellin, cholesterol, phospholipid and carotenoid. We lacked enough material to test for trace constituents but later on with more eggs at our disposal we did and nothing unusual showed up as far as the contents of vitamins, co-enzymes, nucleotides, sulfhydryl groups, et cetera, et cetera were concerned.

One important gross abnormality that showed was the egg's behavior on heating. A small portion of the yolk, heated, "hard-boiled" almost at once. We fed a portion of the hard-boiled egg to a mouse. It survived.

I nibbled at another bit of it. Too small a quantity to taste, really, but it made me sick. Purely psychosomatic, I'm sure.

Boris W. Finley, of the Department of Biochemistry of Temple University—a department consultant —supervised these tests.

He said, referring to the hard-boiling, "The ease with which the egg-proteins are heat-denatured indicates a partial denaturation to begin with and, considering the nature of the shell, the obvious guilt would lie at the door of heavy-metal contamination."

So a portion of the yolk was analyzed for inorganic constituents, and it was found to be high in chloraurate ion, which is a singly-charged ion containing an atom of gold and four of chlorine, the symbol for which is $AuCl_4^-$. (The "Au" symbol for gold comes from the fact that the Latin word for gold is "aurum".) When I say the chloraurate ion content was high, I mean it was 3.2 parts per thousand, or 0.32 per cent.

That's high enough to form insoluble complexes of "gold-protein" which would coagulate easily.

Finley said, "It's obvious this egg cannot hatch. Nor can any other such egg. It is heavy-metal poisoned. Gold may be more glamorous than lead but it is just as poisonous to proteins."

I agreed gloomily, "At least it's safe from decay, too."

"Quite right. No self-respecting bug would live in this chlorauriferous soup."

The final spectrographic analysis of the gold of the shell came in. Virtually pure. The only detectable impurity was iron which amounted to 0.23 per cent of the whole. The iron content of the egg yolk had been twice normal, also. At the moment, however, the matter of the iron was neglected.

One week after Project Goose was begun, an expedition was sent into Texas. Five biochemists went—the accent was still on biochemistry, you see—along with three truckloads of equipment, and a squadron of army personnel. I went along, too, of course.

As soon as we arrived, we cut MacGregor's farm off from the world.

That was a lucky thing, you know—the security measures we took right from the start. The reasoning was wrong, at first, but the results were good.

The Department wanted Project Goose kept quiet at the start simply

because there was always the thought that this might still be an elaborate hoax and we couldn't risk the bad publicity, if it were. And if it weren't a hoax, we couldn't risk the newspaper hounding that would definitely result over any goose-and-golden-egg story.

It was only well after the start of Project Goose, well after our arrival at MacGregor's farm, that the real implications of the matter became clear.

Naturally, MacGregor didn't like the men and equipment settling down all about him. He didn't like being told The Goose was government property. He didn't like having his eggs impounded.

He didn't like it but he agreed to it—if you can call it agreeing when negotiations are being carried on while a machine gun is being assembled in a man's barnyard and ten men, with bayonets fixed, are marching past while the arguing is going on.

He was compensated, of course. What's money to the government?

The Goose didn't like a few things, either—like having blood samples taken. We didn't dare anaesthetize it for fear of doing anything to alter its metabolism, and it took two men to hold it each time. Ever try to hold an angry goose?

The Goose was put under a twenty-four hour guard with the

threat of summary court-martial to any man who let anything happen to it. If any of those soldiers read this article, they may get a sudden glimmering of what was going on. If so, they will probably have the sense to keep shut about it. At least, if they know what's good for them, they will.

The blood of The Goose was put through every test conceivable.

It carried 2 parts per hundred thousand (0.002 per cent) of chloraurate ion. Blood taken from the hepatic vein was richer than the rest, almost 4 parts per hundred thousand.

Finley grunted. "The liver," he said.

We took X rays. On the X ray negative, the liver was a cloudy mass of light gray, lighter than the viscera in its neighborhood, because it stopped more of the X rays, because it contained more gold. The blood vessels showed up lighter than the liver proper and the ovaries were pure white. No X rays got through the ovaries at all.

It made sense and in an early report, Finley stated it as bluntly as possible. Paraphrasing the report, it went, in part:

"The chloraurate ion is secreted by the liver into the blood stream. The ovaries act as a trap for the ion, which is there reduced to metallic gold and deposited as a shell about the developing egg. Relatively high concentrations of unreduced chloraurate ion penetrate the contents of the developing egg.

"There is little doubt that The Goose finds this process useful as a means of getting rid of the gold atoms which, if allowed to accumulate, would undoubtedly poison it. Excretion by eggshell may be novel in the animal kingdom, even unique, but there is no denying that it is keeping The Goose alive.

"Unfortunately, however, the ovary is being locally poisoned to such an extent that few eggs are laid, probably not more than will suffice to get rid of the accumulating gold, and those few eggs are definitely unhatchable."

That was all he said in writing, but to the rest of us, he said, "That leaves one peculiarly embarrassing question."

I knew what it was. We all did.

Where was the gold coming from?

No answer to that for a while, except for some negative evidence. There was no perceptible gold in The Goose's feed, nor were there any gold-bearing pebbles about that it might have swallowed. There was no trace of gold anywhere in the soil of the area and a search of the house and grounds revealed nothing. There were no gold coins, gold jewelry, gold plate, gold watches or gold anything. No one on the farm even had as much as gold fillings in his teeth.

There was Mrs. MacGregor's wedding ring, of course, but she had only had one in her life and she was wearing that one.

So where was the gold coming from?

The beginnings of the answer came on August 16, 1955.

Albert Nevis, of Purdue, was forcing gastric tubes into The Goose —another procedure to which the bird objected strenuously—with the idea of testing the contents of its alimentary canal. It was one of our routine searches for exogenous gold.

Gold *was* found, but only in traces and there was every reason to suppose those traces had accompanied the digestive secretions and were, therefore, endogenous—from within, that is—in origin.

However, something else showed up, or the lack of it, anyway.

I was there when Nevis came into Finley's office in the temporary building we had put up overnight —almost—near the goosepen.

Nevis said, "The Goose is low in bile pigment. Duodenal contents show about none."

Finley frowned and said, "Liver function is probably knocked loop-the-loop because of its gold concentration. It probably isn't secreting bile at all."

"It *is* secreting bile," said Nevis. "Bile acids are present in normal quantity. Near normal, anyway. It's just the bile pigments that are missing. I did a fecal analysis and that was confirmed. No bile pigments."

Let me explain something at this point. Bile acids are steroids secreted by the liver into the bile and *via* that are poured into the upper end of the small intestine. These bile acids are detergentlike molecules which help to emulsify the fat in our diet—or The Goose's—and distribute them in the form of tiny bubbles through the watery intestinal contents. This distribution, or homogenization, if you'd rather, makes it easier for the fat to be digested.

Bile pigments, the substance that was missing in The Goose, are something entirely different. The liver makes them out of hemoglobin, the red oxygen-carrying protein of the blood. Wornout hemoglobin is broken up in the liver, the heme part being split away. The heme is made up of a squarish molecule— called a "porphyrin"—with an iron atom in the center. The liver takes the iron out and stores it for future use, then breaks the squarish molecule that is left. This broken porphyrin is bile pigment. It is colored brownish or greenish—depending on further chemical changes—and is secreted into the bile.

The bile pigments are of no use to the body. They are poured into the bile as waste products. They pass through the intestines and come out with the feces. In fact, the bile pigments are responsible for the color of the feces.

Finley's eyes began to glitter.

Nevis said, "It looks as though porphyrin catabolism isn't following the proper course in the liver. Doesn't it to you?"

It surely did. To me, too.

There was tremendous excitement after that. This was the first metabolic abnormality, not directly involving gold, that had been found in The Goose!

We took a liver biopsy (which means we punched a cylindrical sliver out of The Goose reaching down into the liver.) It hurt The Goose but didn't harm it. We took more blood samples, too.

This time, we isolated hemoglobin from the blood and small quantities of the cytochromes from our liver samples. (The cytochromes are oxidizing enzymes that also contain heme.) We separated out the heme and in acid solution some of it precipitated in the form of a brilliant orange substance. By August 22, 1955, we had 5 micrograms of the compound.

The orange compound was similar to heme, but it was not heme. The iron in heme can be in the form of a doubly charged ferrous ion (Fe^{++}) or a triply charged ferric ion (Fe^{+++}), in which latter case, the compound is called hematin. (Ferrous and ferric, by the way, come from the Latin word for iron, which is "ferrum.")

The orange compound we had separated from heme had the porphyrin portion of the molecule all right, but the metal in the center was gold, to be specific, a triply charged auric ion (Au^{+++}). We called this compound "aureme," which is simply short for "auric heme."

Aureme was the first naturally-occurring gold-containing organic compound ever discovered. Ordinarily, it would rate headline news in the world of biochemistry. But now it was nothing; nothing at all in comparison to the further horizons its mere existence opened up.

The liver, it seemed, was not breaking up the heme to bile pigment. Instead it was converting it to aureme; it was replacing iron with gold. The aureme, in equilibrium with chloraurate ion, entered the blood stream and was carried to the ovaries where the gold was separated out and the porphyrin portion of the molecule disposed of by some as yet unidentified mechanism.

Further analyses showed that 29 per cent of the gold in the blood of The Goose was carried in the plasma in the form of chloraurate ion. The remaining 71 per cent was carried in the red blood corpuscles in the form of "auremoglobin." An attempt was made to feed The Goose traces of radioactive gold so that we could pick up radioactivity in plasma and corpuscles and see how readily the auremoglobin molecules were handled in the ovaries. It seemed to us the auremoglobin should be much more slowly disposed of than the dissolved chloraurate ion in the plasma.

The experiment failed, however, since we detected no radioactivity. We put it down to inexperience since none of us were isotopes men which was too bad since the failure was highly significant, really, and by not realizing it, we lost several weeks.

The auremoglobin was, of course, useless as far as carrying oxygen was concerned, but it only made up about

0.1 per cent of the total hemoglobin of the red blood cells so there was no interference with the respiration of The Goose.

This still left us with the question of where the gold came from and it was Nevis who first made the crucial suggestion.

"Maybe," he said, at a meeting of the group held on the evening of August 25, 1955, "The Goose doesn't replace the iron with gold. Maybe it *changes* the iron to gold."

Before I met Nevis personally that summer, I had known him through his publications—his field is bile chemistry and liver function—and had always considered him a cautious, clear-thinking person. Almost overcautious. One wouldn't consider him capable for a minute of making any such completely ridiculous statement.

It just shows the desperation and demoralization involved in Project Goose.

The desperation was the fact that there was nowhere, literally nowhere, that the gold could come from. The Goose was excreting gold at the rate of 38.9 grams of gold a day and had been doing it over a period of months. That gold had to come from somewhere and, failing that—absolutely failing that—it had to be made from something.

The demoralization that led us to consider the second alternative was due to the mere fact that we were face to face with The Goose That Laid The Golden Eggs; the undeniable GOOSE. With that, everything became possible. All of us were living in a fairy-tale world and all of us reacted to it by losing all sense of reality.

Finley considered the possibility seriously. "Hemoglobin," he said, "enters the liver and a bit of auremoglobin comes out. The gold shell of the eggs has iron as its only impurity. The egg yolk is high in only two things; in gold, of course, and also, somewhat, in iron. It all makes a horrible kind of distorted sense. We're going to need help, men."

We did and it meant a third stage of the investigation. The first stage had consisted of myself alone. The second was the biochemical task-force. The third, the greatest, the most important of all, involved the invasion of the nuclear physicists.

On September 5, 1955, John L. Billings of the University of California arrived. He had some equipment with him and more arrived in the following weeks. More temporary structures were going up. I could see that within a year we would have a whole research institution built about The Goose.

Billings joined our conference the evening of the 5th.

Finley brought him up to date and said, "There are a great many serious problems involved in this iron-to-gold idea. For one thing, the total quantity of iron in The Goose can only be of the order of half a gram, yet nearly 40 grams of gold a day are being manufactured."

Billings had a clear, high-pitched

voice. He said, "There's a worse problem than that. Iron is about at the bottom of the packing fraction curve. Gold is much higher up. To convert a gram of iron to a gram of gold takes just about as much energy as is produced by the fissioning of one gram of U-235."

Finley shrugged. "I'll leave the problem to you."

Billings said, "Let me think about it."

He did more than think. One of the things done was to isolate fresh samples of heme from The Goose, ash it and send the iron oxide to Brookhaven for isotopic analysis. There was no particular reason to do that particular thing. It was just one of a number of individual investigations, but it was the one that brought results.

When the figures came back, Billings choked on them. He said, "There's no Fe^{56}."

"What about the other isotopes?" asked Finley at once.

"All present," said Billings, "in the appropriate relative ratios, but no detectable Fe^{56}."

I'll have to explain again: Iron, as it occurs naturally, is made up of four different isotopes. These isotopes are varieties of atoms that differ from one another in atomic weight. Iron atoms with an atomic weight of 56, or Fe^{56}, makes up 91.6 per cent of all the atoms in iron. The other atoms have atomic weights of 54, 57 and 58.

The iron from the heme of The Goose was made up only of Fe^{54}, Fe^{57} and Fe^{58}. The implication was obvious. Fe^{56} was disappearing while the other isotopes weren't and this meant a nuclear reaction was taking place. A nuclear reaction could take one isotope and leave others be. An ordinary chemical reaction, any chemical reaction at all, would have to dispose of all isotopes equally.

"But it's energically impossible," said Finley.

He was only saying that in mild sarcasm with Billings' initial remark in mind. As biochemists, we knew well enough that many reactions

went on in the body which required an input of energy and that this was taken care of by coupling the energy-demanding reaction with an energy-producing reaction.

However chemical reactions gave off or took up a few kilocalories per mole. Nuclear reactions gave off or took up millions. To supply energy for an energy-demanding nuclear reaction required, therefore, a second, and energy-producing, nuclear reaction.

We didn't see Billings for two days.

When he did come back, it was to say, "See here. The energy-producing reaction must produce just as much energy per nucleon involved as the energy-demanding reaction uses up. If it produces even slightly less, then the overall reaction won't go. If it produces even slightly more, then considering the astronomical number of nucleons involved, the excess energy produced would vaporize The Goose in a fraction of a second."

"So?" said Finley.

"So the number of reactions possible is very limited. I have been able to find only one plausible system. Oxygen-18, if converted to iron-56 will produce enough energy to drive the iron-56 on to gold-197. It's like going down one side of a roller-coaster and then up the other. We'll have to test this."

"How?"

"First, suppose we check the isotopic composition of the oxygen in The Goose."

Oxygen is made up of three stable isotopes, almost all of it O^{16}. O^{18} makes up only one oxygen atom out of 250.

Another blood sample. The water content was distilled off in vacuum and some of it put through a mass spectrograph. There was O^{18} there but only one oxygen atom out of 1300. Fully 80 per cent of the O^{18} we expected wasn't there.

Billings said, "That's corroborative evidence. Oxygen-18 is being used up. It is being supplied constantly in the food and water fed to The Goose, but it is still being used up. Gold-197 is being produced. Iron-56 is one intermediate and since the reaction that uses up iron-56 is faster than the one that produces it, it has no chance to reach significant concentration and isotopic analysis shows its absence."

We weren't satisfied, so we tried again. We kept The Goose on water that had been enriched with O^{18} for a week. Gold production went up almost at once. At the end of a week, it was producing 45.8 grams a while the O^{18} content of its body water was no higher than before.

"There's no doubt about it," said Billings.

He snapped his pencil and stood up. "That Goose is a living nuclear reactor."

The Goose was obviously a mutation.

A mutation suggested radiation among other things and radiation brought up the thought of nuclear tests conducted in 1952 and 1953 several hundred miles away from the site of MacGregor's farm. (If it occurs to you that no nuclear tests have been conducted in Texas, it just shows two things; I'm not telling you everything and you don't know everything.)

I doubt that at any time in the history of the atomic era was background radiation so thoroughly analyzed and the radioactive content of the soil so rigidly sifted.

Back records were studied. It didn't matter how top-secret they were. By this time, Project Goose had the highest priority that had ever existed.

Even weather records were checked in order to follow the behavior of the winds at the time of the nuclear tests.

Two things turned up.

One: The background radiation at the farm was a bit higher than normal. Nothing that could possibly do harm, I hasten to add. There were indications, however, that at the time of the birth of The Goose, the farm had been subjected to the drifting edge of at least two fallouts. Nothing really harmful, I again hasten to add.

Second: The Goose, alone of all geese on the farm, in fact, alone of all living creatures on the farm that could be tested, including the humans, showed no radioactivity at all. Look at it this way: *everything* shows traces of radioactivity; that's what is meant by background radiation. But The Goose showed none.

Finley sent one report on December 6, 1955, which I can paraphrase as follows:

"The Goose is a most extraordinary mutation, born of a high-level radioactivity environment which at once encouraged mutations in general and which made this particular mutation a beneficial one.

"The Goose has enzyme systems capable of catalyzing various nuclear reactions. Whether the enzyme system consists of one enzyme or more than one is not known. Nor is anything known of the nature of the enzymes in question. Nor can any theory be yet advanced as to how an enzyme can catalyze a nuclear reaction, since these involve particular interactions with forces five orders of magnitude higher than those involved in the ordinary chemical reactions commonly catalyzed by enzymes.

"The overall nuclear change is from oxygen-18 to gold-197. The oxygen-18 is plentiful in its environment, being present in significant amount in water and all organic foodstuffs. The gold-197 is excreted via the ovaries. One known intermediate is iron-56 and the fact that auremoglobin is formed in the process leads us to suspect that the enzyme or enzymes involved may have heme as a prosthetic group.

"There has been considerable thought devoted to the value this overall nuclear change might have to the goose. The oxygen-18 does it no harm and the gold-197 is troublesome to be rid of, potentially poisonous, and a cause of its sterility.

Its formation might possibly be a means of avoiding greater danger. This danger—"

But just reading it in the report, friend, makes it all seem so quiet, almost pensive. Actually, I never saw a man come closer to apoplexy and survive than Billings did when he found out about our own radioactive gold experiments which I told you about earlier—the ones in which we detected no radioactivity in the goose, so that we discarded the results as meaningless.

Many times over he asked how we could possibly consider it unimportant that we had lost radioactivity.

"You're like the cub reporter," he said, "who was sent to cover a society wedding and on returning said there was no story because the groom hadn't shown up.

"You fed The Goose radioactive gold and lost it. Not only that you failed to detect any natural radioactivity about The Goose. Any carbon-14. Any potassium-40. And you called it failure."

We started feeding The Goose radioactive isotopes. Cautiously, at first, but before the end of January of 1956 we were shoveling it in.

The Goose remained nonradioactive.

"What it amounts to," said Billings, "is that this enzyme-catalyzed nuclear process of The Goose manages to convert any unstable isotope into a stable isotope.

"Useful," I said.

"Useful? It's a thing of beauty.

It's the perfect defense against the atomic age. Listen, the conversion of oxygen-18 to gold-197 should liberate eight and a fraction positrons per oxygen atom. That means eight and a fraction gamma rays as soon as each positron combines with an electron. No gamma rays either. The Goose must be able to absorb gamma rays harmlessly."

We irradiated The Goose with gamma rays. As the level rose, The Goose developed a slight fever and we quit in panic. It was just fever, though, not radiation sickness. A day passed, the fever subsided, and The Goose was as good as new.

"Do you see what we've got?" demanded Billings.

"A scientific marvel," said Finley.

"Man, don't you see the practical applications? If we could find out the mechanism and duplicate it in the test tube, we've got a perfect method of radioactive ash disposal. The most important drawback preventing us from going ahead with a full-scale atomic economy is the headache of what to do with the radioactive isotopes manufactured in the process. Sift them through an enzyme preparation in large vats and that would be it.

"Find out the mechanism, gentlemen, and you can stop worrying about fallouts. We would find a protection against radiation sickness.

"Alter the mechanism somehow and we can have Geese excreting any element needed. How about uranium-235 eggshells?

"The mechanism! The mechanism!"

We sat there, all of us, staring at The Goose.

If only the eggs would hatch. If only we could get a tribe of nuclear-reactor Geese.

"It must have happened before," said Finley. "The legends of such Geese must have started somehow."

"Do you want to wait?" asked Billings.

If we had a gaggle of such Geese, we could begin taking a few apart. We could study its ovaries. We could prepare tissue slices and tissue homogenates.

That might not do any good. The tissue of a liver biopsy did not react with oxygen-18 under any conditions we tried.

But then we might perfuse an intact liver. We might study intact embryos, watch for one to develop the mechanism.

But with only one Goose, we could do none of that.

We don't dare kill The Goose That Lays The Golden Eggs.

The secret was in the liver of that fat Goose.

Liver of fat goose! *Pate de foie gras!* No delicacy to us!

Nevis said, thoughtfully, "We need an idea. Some radical departure. Some crucial thought."

"Saying it won't bring it," said Billings despondently.

And in a miserable attempt at a joke, I said, "We could advertise in the newspapers," and that gave *me* an idea.

"Science fiction!" I said.

"What?" said Finley.

"Look, science-fiction magazines print gag articles. The readers consider it fun. They're interested." I told them about the thiotimoline articles Asimov wrote and which I had once read.

The atmosphere was cold with disapproval.

"We won't even be breaking security regulations," I said, "because no one will believe it." I told them about the time in 1944 when Cleve Cartmill wrote a story describing the atom bomb one year early and the F.B.I. kept its temper.

"And science-fiction readers have ideas. Don't underrate them. Even if they think it's a gag article, they'll send their notions in to the editor. And since we have no ideas of our own; since we're up a dead-end street, what can we lose?"

They still didn't buy it.

So I said, "And you know— The Goose won't live forever."

That did it, somehow.

We had to convince Washington; then I got in touch with John Campbell and he got in touch with Asimov.

Now the article is done. I've read it, I approve, and I urge you all not to believe it. Please don't.

Only—

Any ideas?

THE END

⋆14⋆

Temporal Chirality*
The Burgenstock Communication

Brian Addle[1], Michael J. S. Dewar, Hans Krapp[2], Till E. Spiegel[3], and (in part) Michel Obb[4]

Editor's Preface

The composing of spoofs has a long and venerable history, treating broad areas as well as narrow specialties. Such spoofs can reflect the professional at play and may, in some cases, be informally classified as speculative fiction. To enjoy some spoofs may require a level of professional expertise in a very specialized area. For those of us unfamiliar with the specialized area involved, the cleverness of any such offering is lost; for those of us in familiar territory, such offerings are a unique delight. The compositions have occasionally been prepared, usually as in-group jokes,

1. Present address: Egg Research Institute, Poultry Lane, Birdwell, Yolkshire, England.
2. Present address: Chiral Research Unit, Department of Defense (CRUDD), Washington, DC 20301.
3. Present address: Institut fur Schöpferisch Wissenschaftliches Denken, Universität, Munchausen 2, West Germany.
4. AA Predoctoral Fellow.

by some of our most distinguished colleagues, in this case Professor Michael J. S. Dewar. While it is representative, it is at the more complex end of the genre's spectrum.

★　★　★　★

★ It has been generally believed that optical activity is a property shown only by chiral molecules or crystals. We wish to report some observations which suggest that this view may ★ need to be modified.

In discussing the chirality of molecules we consider them to have their equilibrium geometries. In practice, however, a molecule has vibrational energy, even at 0 K, and the vibrations will instantaneously destroy the symmetry responsible for the lack of chirality. Thus while ethylene ($H_2C=CH_2$) is a planar and achiral of equilibrium, having D_{2h} symmetry, a rotation of one methylene group relative to the other through any angle less than 90° reduces the symmetry to C_{2v} and so makes it chiral. Such chirality is of course reversed during the course of the corresponding vibration, and the molecule on average possesses D_{2h} symmetry. In the case of a single vibration, dextro- or levochirality would persist only one-half of the vibrational period, i.e. ~0.1 psec. It occurred to us, however, that in the case of a larger molecule with many normal modes, a kind of beat phenomenon might lead to the persistence of chirality for much longer periods of time. If such a compound were then generated suddenly by a chiral process, it might show bulk optical activity long enough for this to be observed. Such a process could be a photochemical reaction brought about by circularly polarized light. We accordingly devised the following simple experiment.

Photolysis of diazodeoxybenzoin **1** gives diphenylketene **3** via the carbene **2**. Using a mode-locked ruby laser with a frequency doubler and quarter wave plate, we were able to generate picosecond pulses of circularly polarized light of a frequency sufficient to bring about the conversion of **1** to **3**. The reaction was carried out in solution, any optical activity being detected by a passing light from a second laser through a polarizer, the solution, a second polarizer orthogonal to the first, and a photomultiplier detector. The output from the detector was displayed on an oscilloscope with a very fast sweep (1 nsec), triggered to start at the instant of the photolytic flash.

$$Ph-\underset{\underset{\displaystyle N_2}{\|}}{C}-CO-Ph \xrightarrow[-N_2]{h\nu} Ph-\ddot{C}-CO-Ph \rightarrow Ph_2C=C=C$$

$$\mathbf{1} \qquad\qquad\qquad \mathbf{2} \qquad\qquad\qquad \mathbf{3}$$

We hoped to observe a pulse in the oscilloscope trace corresponding to momentary chirality of the reaction product. We were at first disappointed to see only a horizontal line. However, we then realized that the line was displaced from the position corresponding to zero signal. Not only had chirality developed, but it had not decayed detectably during the 1-nsec sweep! We therefore progressively slowed the sweep by successive factors of ten. To our amazement no change occurred until it had been slowed to 100 μsec, when decay of the signal at last became visible. At still slower sweep rates the trace became oscillatory due to periodic fluctuations of chirality with molecular vibrations. Analysis of the signal with a spectrum analyzer gave a broad peak of frequencies centered on 1.26 kHz, this being clearly the principal chiral frequency of **3**.

In view of this astonishing result we examined the theory of this phenomenon, which we term "temporal chirality". It soon became clear that any ab initio treatment would be out of the question. Fortunately, however, a new and very powerful semiempirical SCF MO treatment (MINDO/5b) had just been developed here; using this and perturbation theory we were able to arrive at a simple expression for the principal chiral frequencies of molecules, the calculated value for **3** agreeing with experiment to within 17%. According to this theory, the chiral lifetimes of molecules should vary as the cube of the number of atoms in them. The chiral life times of large molecules, such as proteins or nucleic acids, should therefore extend to hours, or even days.

This conclusion clearly has some fascinating consequences. It could, for example, provide a simple explanation of the chirality of living matter; for long-lived temporal chirality of otherwise inactive species could lead to induced optical activity in the products of reactions undergone by them. Similar syntheses of optically active materials from inactive reagents could well be observed in the laboratory, and one such reaction has indeed been claimed (*1*). Some of the anomalous results recently reported in attempts to determine the absolute configurations of molecules by X-ray crystallography could also well have been due to chiral fluctuations. Since a crystal can, in the present connection, be regarded as a single molecule, temporal chirality in a crystal could persist for long periods. We suggest that in the future only very small crystals be used in such studies, and that they should first be allowed to relax in a dark achiral environment for at least six months.

Some even more preliminary studies have indicated that chiral oscillations may be observed in thermal reactions in the gas phase if carried out in a properly tuned cavity. This suggests the possible use of such oscillators (*chiralasers*) as frequency standards in communication systems. It is interesting in this connection that a marked difference seems to exist between the chiral frequency spectra of pericyclic reactions depending on whether or not they obey the Woodward-Hoffmann rules. The signals from "allowed" reactions contain simple harmonic progressions and indeed, as Mr. Obb has ingeniously shown, produce melodious sounds when fed into an amplified speaker system. The "forbidden" reactions on the other hand have very complex spectra and give rise to most unpleasant sounds. This result suggests that music may also be subject to principles analogous to the Woodward-Hoffmann rules, much modern music being "forbidden" in this sense.

However, a note of warning should be struck. Apart from the speculative nature of some of our suggestions, it must be noted that while the experiments reported here were carried out exactly a year ago, and a preliminary account was given at the Burgenstock Conference on Stereochemistry on May 4, 1973, circumstances have so far prevented us from repeating them. We had indeed delayed publication in the hopes that we might be able to remedy the deficiency; since this now seems unlikely, we are reporting our results in the hope of stimulating further work elsewhere.

Reference

1.   K. Paranjape, N. L. Phalniker, B. V. Bhide, and K. S. Nargund. *Current Science* 12, 150 (1943); *Nature*, 153, 141 (1944).

Editor's Addendum

Professor Dewar offered the following supplementary information (personal communication).

This paper was presented at an annual Burgenstock Conference on Stereochemistry held in Switzerland in 1973. The conference had a tradition for someone to present a "frivolous" lecture as part of the regular program. Since the lectures were strictly by invitation, only those who had attended a previous conference even knew of the tradition and only the perpetrator knew in advance which lecture. As a result, members of the very high-level audience took varying lengths of time to recognize the fraud, and it appears that one of them subsequently asked a serious question (while the rest of the audience was openly laughing).

Isn't that a marvelous tradition? Must serious chemists also be solemn?

Encouraging Creativity
in the Classroom

★15★

Science Fiction

A Classroom Resource

Jack H. Stocker

★Throughout modern times, selected science fiction has enjoyed a respected place in academia. Almost exclusively utilized in the liberal arts areas, such titles as *Brave New World*, *1984*, and *Looking Backward* are immediately recognized. More recently these have been supplemented, e.g., by *Fahrenheit 451* (Bradbury), *The Dispossessed* (LeGuin), *Lord of the Flies* (Golding), *The Handmaid's Tale* (Atwood) and *Slaughterhouse Five* (Vonnegut) among many others. The choices tended to reflect the social sciences and, in general, paid little attention to any traditional science involved except as a background component. They would not be considered to be "science-driven" and the *teaching* of science fiction has largely been a province of English departments. (As a case in point, permission for the author to teach a non-credit course specifically entitled "Science Fiction" required the permission of his university's English department while the latter's teaching of a credit course identically entitled did not require a corresponding clearance from our college of sciences.) Nonetheless, an enormous array of science fiction courses are offered throughout the country. Some 25 years ago, the highly respected SF writer Jack Williamson, an academician whose dissertation dealt with H. G. Wells, solicited nationwide

233

input describing science fiction courses then being offered; his subsequent compilation showed approximately 140 taught at all levels, to a wide array of audiences, and reflecting a variety of approaches and contents. The number of such courses can currently be confidently extrapolated to be in the thousands and growing. While there is a significant amount of published material purporting to offer classroom guidance, a single volume, edited by Jack Williamson and entitled *Teaching Science Fiction—Education for Tomorrow* (Olwswick Press, Philadelphia, 1980) can be recommended as perhaps the most useful single reference. Among some 25 chapters, many by leading writers in the field, there is one by Stanley Schmidt (presently the editor of *Analog Science-Fact and Science-Fiction* as well as a university professor of physics) entitled *Science Fiction and the Science Teacher*. He describes his experiences in the classroom teaching *science* via science fiction. The book, in general, can be highly recommended.

It might be useful to point out that exploiting SF in the classroom can be approached in two different ways. One, the more traditional, utilizes the science fiction available in the various media (predominantly prose) and has the reader/viewer evaluate it, i.e., analyze it for content. The alternate approach invites the student to create a fantasy, i.e., to write a story or essay to provide a "creative writing" response answering a chemistry assignment. A closely related but more sophisticated approach involves "molecular anthropomorphism". These two creative approaches, with references, are considered in somewhat more detail in this chapter.

The following is an attempt to offer some nuts-and-bolts guidance on how science fiction can be utilized in the classroom. Much of it was originally presented as part of the symposium talk that became Chapter 1 and follows directly from the material in that chapter.

First, the Problem

1. Probably little can be done where class size is large, limiting its use in freshman classes at larger universities. (But see L. Miller's publication, described later in this chapter, where a creative writing technique has been used successfully with classes as large as 100–200 students.)

2. Clearly required is a commitment and a knowledgeability on the part of the teacher. While the former does not present a problem, the latter does unless there is a formal lesson plan and a specific text or reference material with which a teacher can familiarize himself or herself.

3. It should be recognized that the SF orientation many students have is not to the "hard science" category but to the writings of Edgar Rice Burroughs, Stephen King, or J. R. R. Tolkien. This is clearly a frail support on which to build a dialogue. The approach must involve stressing that understanding the valid/invalid premises on which the SF or pseudo-SF story is predicated would *increase* the pleasure to be derived from the writing. It does not spoil a story or poem to analyze it. A lazy mind is thereby stretched and the way is opened to considerably enhanced reading pleasure.

Where Can Ancillary Support Be Found?

1. **School library holdings.** Libraries vary in their receptivity to having a unique niche for SF. A great deal of quality used material can be purchased cheaply; clearly this must be done by someone who knows the good from the bad. The school PTA or even the library itself might be willing to spring for perhaps $50 for the "knowledgeable person" to spend at local thrift shops, church sales, used-book stores, and similar money-saving outlets on the understanding that the books procured (hardcover and softcover) would establish an "SF corner". The possibility of library video holdings is particularly attractive. Where copyright laws permit, non-profit reproduction could be utilized. TV can be even more effective than books.

2. **School bookstores can be utilized** where multiple copies of a very few paperbacks are wanted. The problem is that *collections* of short stories, much preferred, have poorer density of wanted material. Most good SF remains in print and is available. Again, a knowledgeable mentor is needed.

3. **Student organizations.** If the school does not already have a SF club, see if there is any support for one. (Support is traditionally strongest in the physics area). Some organizations, e.g., the New England Science Fiction Association (NESFA), originally based at MIT, are sufficiently well established that they are incorporated and even serve as publishing houses.

4. **Science fiction lesson units** are often available from the appropriate resource departments. Check with your state, local and university departments of education.

How Can These Ancillary Resources Be Utilized?

1. **By the Teacher.** Where classes are large and SF club interest is apathetic and the library is rigid in its traditional pattern, etc., there is still room for utilization of SF. The teacher can use an occasional plot twist to illustrate currently presented material and enliven lectures. A few examples:

    A. *Antimatter.* A communication officer aboard a starship is in communication with his counterpart as his ship approaches her star system. Over a period of time they fall deeply in love. Only when the starship approaches her system is it realized that her sun and planets consist of antimatter (inside-out matter, Contra-Terrene matter, C.T. or Seetee matter) and that any physical contact between the two lovers would produce their total annihilation in a blast of hard radiation. Here is the ultimate story of Romeo and Juliet, the doomed lovers.

    B. *Chiral opposites.* What would be the consequences of encountering a world where evolutionary processes had produced mirror-image/unnatural sugars and amino acids. Foods would look the same, but would not taste or smell the same, or *serve the same nutritional roles.* Would the consumer of the proteins and carbohydrates formed from these "unnatural" building blocks starve to death on such a planet? Biochemical phenomena and some quite grandiose concepts such as "chiral universes" can be mentioned. Various authors have explored some aspects of this concept.

2. **By Individual Students.** Formally assign projects (film, TV or book evaluation) to specific students to be presented as written reports or as oral presentation to their class for supplemental credits. Or, set aside perhaps 10 minutes at the beginning or end of the class period once a week for any student to report verbally on any interesting SF chemistry they've encountered in their reading or viewing, with brief class response to the report.

3. **By the Entire Class.** Include an occasional bonus question on an examination, inviting students to speculate on the consequence of a change in some natural phenomena, i.e., an artificial "factoid". For example, what if ice was more dense than water? One could, of course, assign projects as in (2) above to an entire class.

An Alternative Approach: Creative Writing

Dr. Naola Van Orden (a former teacher at Sacramento City College) has described her classroom-based research in a report entitled "Is Writing an Effective Way to Learn Chemical Concepts?" (*Chem. Educ.* **1990**, *67*, 583). While her experiment yielded mixed results, the details offer interesting and useful information. She has offered a short paper (ibid., p. 1052), provocatively entitled "Once upon a Time in the Land of Chemistry", subtitled "A Case for Fantasy Writing in Chemistry". An example of her approach is provided by the following assignment.

> Write a fantasy story in which a group of acetate ions were visited by a group of hydrogen ions and later by a group of hydrochlorite ions. Use Le Chatelier's principle and a table of ionization constants to help you decide what will happen. (If you really want to get involved, pretend you are one of the characters in your story.) After you finish the story, explain in your own words, with chemical equations, what happens when NaClO is added to a solution of $HC_2H_3O_2$.

A representative response anthropomorphisized the ions involved in the form of a letter sent by one acetate ion to her acetate friend, Agnes. It seems that the former's party of four acetates met up with "the cutest bunch of hydrogens", paired off and were getting along famously when an older sister "chlorous" and her friends showed up and promptly stole the hydrogens away. The acetate ion sadly provided a moral: "not to bond with just any ion (you) meet. While they may act positive to you,... as soon as they meet someone who seems to need them more, off they go."

In a somewhat different approach, the student may be asked to write an autobiography of an element:

> Select a chemical, look up its properties in a handbook and in your text and then describe those properties by giving them anthropomorphic qualities. Write as if you are the element. (Van Orden, personal communication).

A related but more sophisticated use of fantasy has been reported by Miller (*J. Chem. Educ.* **1992**, *69*, 141). Organic chemistry classes at the University of Minnesota have been asked to write essays on "molecular anthropomorphism" using the following guidelines.

> Think about molecules as people. Think about a chemical reaction between two molecules as an interaction between two humans. Think

about molecular structure as human motivation. Write a metaphor in which molecules take on human characteristics.

For example: A Lewis acid and a base bond together. The reaction occurs because the base wants to donate electrons and the acid needs more electrons. By bonding together they satisfy their needs, and they are stabilized. The structural element that motivates the base is its unshared pair of electrons.

In a metaphor you might make the Lewis base a man and the acid a woman. The reaction is then marriage. They bond. The motivation could be sex, money, or some other quality they can share. If money becomes the metaphoric equivalent of electrons, the metaphor can be developed by assuming that rich men (strong bases) are more attracted to women (acids) than poor men. Indeed, chemists often speak of strong bases as being electron-rich. With this metaphor the reactions of various acids and bases can be compared.

The author points out that we already employ a number of metaphors in chemistry that impute emotional content to molecules, e.g., they *share* electrons, are hydro*phobic*, or nucleo*philic*, or in *excited*, *perturbed*, or *degenerate* states. Strong emphasis is placed on the quality of the creativity in the writing assignment.

The resulting essays, stories, poems, song lyrics, newspaper reports, comic books, and even one elaborate picture indicated that this was a successful "writing to learn" exercise. (The papers were not considered to be good literature.) However, the chemistry was almost always accurate, and it was often extensive and clearly described.

Both of the above authors provide grading guidance in the references given.

Obviously any creative writing approach must enjoy departmental support to be effective. There is clearly a vast amount of little-explored territory here. The creative writing approach, it can be argued, makes Chemistry more accessible and user-friendly, both of which are major goals being aggressively pursued. It deserves further exploration.

The January 26, 1998 issue of *Chemical and Engineering News*, p 65, carried the report of an ACS Award to Dr. Zafra Lerman for "Encouraging Disadvantaged Students into Careers in the Chemical Sciences." She is credited with raising the number of undergraduates taking science courses at Columbia College (a small liberal arts college near Chicago) from 20 to 2,000 during her tenure and for conducting courses and workshops in science teaching at all levels and for the general public. She has shown an effective "ability to incorporate science and math with students' hobbies, interests, and cultural backgrounds (that) has resulted in student projects ranging from paintings and sculptures to songs and dances."

> A group of theater students, for example, presented a science fiction script, "Sustaining Life on the Planet Zafra," in which scientists are kidnapped and taken to the planet, which has lost its sun. The scientists are forced to explain how to use heat from the core of the planet as an energy source.

> Another group of literature students created a spoof of "Romeo and Juliet" titled "Sodium and Chlorine: A Love Story." While telling the romantic story of the bonding of these two elements, the actors manage to provide descriptions and explanations of chemical bonding, ionic solution, metals versus nonmetals, and periodicity, among other concepts.

> A script created by a dance group, "Shield or Tragedy of the Skies," describes the formation and depletion of Earth's ozone layer. Groups of three dancers represent ozone molecules, and dancers in pairs represent oxygen. "Chlorofluorocarbons" dance menacingly near the ozone layer.

A Bottom Line

If students sense that a teacher will listen sympathetically to their SF enthusiasms, they will come to him or her to share them, thereby providing the teacher with an invaluable opportunity to communicate with a receptive minds. *That*, of course, is what teaching is all about.

∗16∗

Using Science Fiction To Help Teach Science

A Survey of Chemists and Physicists

*Clarence J. Murphy, Mary Ann Mogus,
and Patricia M. Crotty*

★Promoting an interest in science is a weak point of the present educational system in the United States, despite the integral role that science and technology play in American society. Our shared concerns and interest in science fiction inspired us to undertake a project in which academic chemists and physicists would be surveyed to determine the feasibility of using science fiction as a supplementary tool in the teaching of science. In addition, the survey would be used to solicit examples of books, films, and television programs for a multimedia bibliography that could be a source of supplementary materials for the teaching of science.

The use of science fiction in the teaching of science was investigated by Sandery (1), Schmidt (2), and more recently by Barra (3). Isaac Asimov (4) also made suggestions for the use of science fiction in the teaching of science. Leroy Dubeck (5) in *Science in Cinema* attempted to provide curriculum ideas.

These authors suggested that there is support for the use of science fiction as a means of sparking student interest in science. To determine if support indeed exists in the academic community, we developed and distributed a survey to a random sample of academic chemists and physicists. This chapter presents the results of that survey and a bibliography of suggested science fiction books, films, and television programs.

How We Went About It

The survey populations were randomly chosen from *College Chemistry Faculties* (6) and the *American Physical Society Membership Directory* (7). The survey responses were analyzed by percent difference and chi-square tests of significance at an accuracy level of 95% significance with a ±6% tolerated error. The minimum number of responses to achieve this significance level is 267, which was exceeded for both chemists and physicists. A mail survey was used and, because a 30% response rate was expected, the survey was sent to 1000 individuals chosen from among the academic members listed in the directories.

Each survey form was given a code number so that respondents could be compared with a master list when the form was returned. This procedure was used to identify respondents in case a second mailing was required and to identify those who expressed an interest in the use of science fiction as a teaching tool. The survey was also administered by telephone to a random sample of those who had not replied by mail to determine if their opinions differed from those who did return the mail survey. Significant differences were not apparent between the two groups. Complete usable survey forms were obtained for 277 chemists and 290 physicists.

The purpose of the survey and the manner in which the results would be used was explained in a cover letter. The survey form included a number of demographic questions that were selected to categorize the respondents by age, sex, whether they taught, and specialization within chemistry or physics. The main part of the survey consisted of nine statements for which the respondents were given the choices: strongly agree, agree, neutral, disagree, or strongly disagree. The statements were as follows:

1. The value of science fiction lies primarily in escapism or fantasy.

2. Science fiction has a predictive value. For example, it can foretell developments in the sciences.

3. Science fiction portrays possible future events.

4. Science fiction portrays science in a positive fashion.

5. Science fiction portrays scientists in a negative fashion.

6. Science fiction frequently abuses known scientific facts.

7. Science fiction fosters interest in science.

8. Science fiction should adhere to known scientific laws.

9. Science fiction influences how the public views science.

The respondents were also asked to recommend science fiction works that they considered valuable to enable us to establish a foundation on which to develop a list of science fiction works that could be used by teachers.

The Responses

Who Were the Respondents?

The demographic questions offered an opportunity to study the composition of the academic chemistry and physics communities. The results are shown in Table 1.

Table 1. Demographics of Respondents

Characteristic	Chemists		Physicists	
	Men	Women	Men	Women
Number	247	34	272	36
Percent	87.9%	12.1%	88.8%	11.2%
Age <45	40.2%	44.2%	44.1%	77.8%
Age >45	59.8%	55.8%	55.9%	22.2%
Teachers				
age <45	98.7%	93.3%	33.3%	32.1%
age >45	99.4%	100%	48.1%	50%
Not teachers				
age <45	1.3%	6.7%	66.7%	67.9%
age >45	0.6%	0%	51.9%	50%
References in class				
age <45	19.2%	14.2%	32.5%[a]	25%[a]
age >45	25.7%	31.5%	23.3%[b]	0%[b]

[a]Sample includes 40 men and 9 women.

[b]Sample includes 73 men and 4 women.

This portion of the survey revealed several interesting results. As expected, the members of both professions were overwhelmingly male, and a substantial majority of the men were over age 45. Surprisingly, the proportion of women chemists over 45 was more than twice as great as the proportion of women physicists. This situation may reflect the fact that women have penetrated academic physics departments much later than chemistry departments. Also quite unexpected was the response from more than half of the physicists stating that they do not teach. This fact in part may reflect the different compositions of the chemists' and physicists' databases. An inspection of the addresses of the chemists showed that almost all are in either academic chemistry, biochemistry, or chemical engineering departments. On the other hand, the addresses of the physicists indicated that a substantial number are affiliated with various kinds of academic research facilities.

Their Opinions

Responses to the outliers were minimal, and because combining data would simplify analysis, strongly agree and agree, and strongly disagree and disagree were combined for the purpose of chi-square analysis. The results are as follows.

Statement 1.
The value of science fiction lies primarily in escapism or fantasy. Chemists agree if they refer to science fiction in class ($p < 0.01$). Physicists show no pattern.

Statement 2.
Science fiction has a predictive value. For example, it can foretell future developments in science. Chemists agree if they refer to science fiction in class ($p < 0.01$) or watch science fiction TV programs ($p < 0.01$). Physicists agree if they read science fiction books ($p < 0.01$).

Statement 3.
Science fiction portrays possible future events. Chemists agree if they watch science fiction films ($p < 0.01$) or TV programs ($p < 0.01$). Physicists agree if they refer to science fiction in class ($p > 0.05$), read science fiction books ($p < 0.01$), or watch science fiction films ($p < 0.01$) or TV programs ($p < 0.01$).

Statement 4.
Science fiction portrays science in a positive fashion. Both chemists and physicists agree if they read science fiction books ($p < 0.01$) or watch science fiction films ($p < 0.01$) or TV programs.

Statement 5.
Science fiction portrays scientists in a negative fashion. Chemists disagree if they read science fiction books ($p < 0.01$) or watch science fiction films ($p > 0.05$) or TV programs ($p > 0.05$). Physicists disagree if they read science fiction books ($p < 0.01$).

Statement 6.
Science fiction frequently abuses known scientific facts. Chemists disagree if they read science fiction books ($p < 0.01$). Physicists show no pattern.

Statement 7.
Science fiction fosters an interest in science. Both chemists and physicists agree if they read science fiction books ($p < 0.01$) or watch science fiction films ($p < 0.01$), or TV programs ($p < 0.01$).

Statement 8.
Science fiction should adhere to known scientific laws. Chemists over 45 agree ($p < 0.01$), and under 45 disagree ($p < 0.01$) if they watch science fiction films. Physicists disagree if they read science fiction books ($p < 0.01$) or watch science fiction films ($p < 0.01$) or TV programs ($p < 0.01$).

Statement 9.
Science fiction influences how the public views science. Chemists show no pattern. Physicists agree if they read science fiction books ($p > 0.05$).

Because innovation in teaching requires the active cooperation of faculty, positive responses to Statements 4 and 7 are crucial. Chemists and physicists who are familiar with science fiction materials agree with both of these statements or see them as neutral well within the tolerated error limits of the survey. Thus, those who are familiar with science fiction may be willing to consider it as an instructional supplement.

Response differences between chemists and physicists are apparent for Statements 1 through 3 and 5 through 9. These differences appear to be media-dependent and, at least for chemists responding to Statement 8, age-dependent.

For chemists both age and whether they refer to science fiction in class (26.6% claim to do so to some extent) affect their responses to Statements 1 and 8.

Academic scientists must have a positive image of science fiction if they are to be induced to use it as a supplement in the teaching of science. The results of the survey show that scientists familiar with science fiction in some form perceive that science fiction portrays science in a positive manner. These scientists also believe that science fiction has a predictive value for possible future events and developments in science. Both chemists and physicists agree that an interest in science is fostered by science fiction.

The presentation of scientific laws in science fiction is viewed in a flexible manner by those familiar with science fiction. Only chemists over 45 agree that adherence to known scientific laws is necessary. This attitude bodes well for the acceptance of science fiction as a supplement to the teaching of chemistry and physics because science fiction is frequently set within a highly speculative scientific framework.

An interesting and unexpected result is evident from the data in the opinion statements. Where a pattern is seen in the responses of physicists, a familiarity with science fiction through reading is always present ($p < 0.01$). No such pattern is discernible for chemists. This result suggests that for a given discipline, the successful introduction of science fiction as a teaching supplement may be medium-dependent, at least for the instructor.

The Recommendations

A total of 311 individuals, 55.8% of the total number of respondents, submitted examples of recommended science fiction. The science fiction media preferences by age are shown in Table 2. The science fiction books, films, and television programs that received four or more recommendations by chemists are listed in Table 3.

The data in Table 2 indicate that media preference may also be age-dependent. Unexpectedly, for both chemists and physicists under 45 the proportion of recommended science fiction works is heavily weighted toward books. On the other hand, for chemists and physicists over 45 the proportions of recommendations for books and films are almost equal. In all cases less than 20% of the science fiction citations are to television programs, and the overwhelming majority of these recommendations are for *Star Trek* and *Star Trek: The Next Generation*.

Table 2. Media Preferences of Respondents by Age Group

Media	Chemists		Physicists	
	Under 45	Over 45	Under 45	Over 45
Books	50.2%	39.9%	49.6%	40.5%
Films	33.5%	39.0%	32.4%	41.9%
TV	16.3%	21.3%	18.0%	17.6%

Note: For chemists under 45, there were 52 replies, 257 citations, and 4.94 citations per response. For chemists over 45, there were 78 replies, 272 citations, and 3.49 citations per response. For physicists under 45, there were 87 replies, 377 citations, and 4.33 citations per response. For physicists over 45, there were 64 replies, 227 citations, and 3.55 citations per response.

Table 3. Television Programs, Movies, and Books Recommended by Respondents

Title	Chemists	Physicists
Television Programs		
Star Trek	50	51
Star Trek: The Next Generation	25	29
Twilight Zone	7	5
Dr. Who	5	10
Outer Limits		4
Hitchhiker's Guide to the Galaxy		4
Other titles	9, 1–2 each	5, 1 each
Movies		
Star Wars	30	37
2001: A Space Odyssey	26	37
Star Trek 1	20	20
The Empire Strikes Back	13	12
The Return of the Jedi	13	
Star Trek 2	10	7
Star Trek 4	10	4
Forbidden Planet	10	13
Alien		12
Aliens		11
Star Trek 3		8
ET: The Extraterrestrial		7
Bladerunner		6
The War of the Worlds		5
Close Encounters of the Third Kind		4
Total Recall		4
Other films	41, 1–3 each	38, 1–3 each

Continued on next page

Table 3. Television Programs, Movies, and Books Recommended by Respondents—*Continued*

Title	Chemists	Physicists
Books		
Foundation Series (Asimov)	24	20
Dune (Herbert)	13	10
The Lord of the Rings (Tolkien)	11	
Robot series (Asimov)	10	
2001: A Space Odyssey (Clarke)	9	7
Dune series (Herbert)	7	4
The Andromeda Strain (Crichton)	7	
Stranger in a Strange Land (Heinlein)	6	8
The Time Machine (Wells)	6	
The Martian Chronicles (Bradbury)	6	6
The War of the Worlds (Wells)	5	
20,000 Leagues Under the Sea (Verne)	4	6
Fantastic Voyage (Asimov)	4	
Ringworld (Nevin)		7
The Mote in God's Eye (Nevin/ Pournelle)		7
Childhood's End (Clarke)		6
Hitchhiker's Guide to the Galaxy (Adams)		5
Ringworld series (Nevin)		4
Fahrenheit 451 (Bradbury)		4
A Canticle for Leibowitz (Miller)		4
Robot (Asimov)		4
Other titles	85, 1–3 each	87, 1–3 each

Similarly, the great majority of film recommendations are for either *2001: A Space Odyssey*, or one of the four *Star Trek* films. *Forbidden Planet*, *Alien*, and *Aliens* also received a substantial number of recommendations.

The most cited book authors are Isaac Asimov, Ray Bradbury, Frank Herbert, Arthur C. Clarke, Robert Heinlein, J. R. R. Tolkien, H. G. Wells, and Jules Verne. The most cited books are the *Foundation* series of Isaac Asimov and the *Dune* series of Frank Herbert.

Summing Up

The data from these surveys indicate that a substantial number of academic chemists and physicists would be receptive to the use of science fiction as a supplement in the teaching of science. In fact, approximately 25% now make at least some reference to science fiction in their teaching. The data also suggest that there may be a difference in media preference for the different disciplines. Curricula suggestions should thus include a variety of different media presentations of science fiction, so that teachers may choose according to their preferences and familiarity.

Acknowledgments

We acknowledge the assistance of Jack Swineford for the computer analysis of the data and Constance G. Murphy for the compilation of the demographic data. We acknowledge the students of P. M. Crotty's political science class who conducted the telephone survey as their practical introduction to telephone polling. We also express our sincere appreciation to the more than 600 of our colleagues who took the time to complete the survey.

References

1. Sandery, P. "Science Fiction and Legitimate Science", *South Aust. Sci. Teach. J.* **1973**, *73*, 35.
2. Schmidt, S. A. "Science Fiction Courses: An Example of Some Alternatives", *Am. J. Phys.* **1973**, *41*, 1052.
3. Barra, P. A. "It Was a Dark and Stormy Chemistry Class", *Sci. Teach.* **1988**, *55*(7), 33.
4. Asimov, Isaac "It's Such a Beautiful Day," *Social Educ.* **1973**, *37*, 112.
5. Dubeck, L. W. *Science in Cinema*; Teachers College Press: New York, 1988.
6. *College Chemistry Faculties*; 8th ed.; American Chemical Society: Washington, DC, 1989.
7. Bulletin of the American Physical Society Membership Directory; Vol. 35, 1991.

★ 17 ★

Space, Time, and Education

John E. Arnold

★A unique type of course in creative thinking is in operation at the Massachusetts Institute of Technolgy—science fiction as a laboratory technology!

Professor John E. Arnold, in this article, gives one of the first discussions of the unique, and highly interesting educational technique he and his co-workers have developed at M.I.T.

Most of the articles we run in this magazine have to do with developments of physical science; there are very, very few social-science inventions available for discussion, wherein a clean-cut break-away from traditional methods can be defined concretely, the reasons for the break-away stated clearly, and the theory behind the change made definite.

Yet in our present world of gadgets, machines, and highly developed physical technology, social inventions are the crying need of Mankind. Perhaps a major reason for the extreme paucity of social invention is the lack of just such training in creative thinking as Professor Arnold's course is specifically designed to provide.

There is a curious and confusing paradox in the nature of human progress; men have, down the ages, been willing to fight and die for the ideals they hold valid and important. Men have shown full willingness to total self-sacrifice in defense of their heritage.

Yet by the very meaning of the concepts, it is impossible, and forever will be impossible, to maintain the "Ancient Heritage" and progress in any way! No man today can defend the democracy that Washington and Jefferson established, because America has developed, has learned greater wisdom and invented new social ideas, the "heritage" of Washington and Jefferson is forever gone!

For example, in their day, their concept of democracy held that no man who owned less than five thousand dollars worth of property had a right to vote. Their concept of democracy has long since been changed; they would never have accepted the idea of woman voters.

The very fact that men are idealists, and will fight for their ideals, makes social inventions extremely difficult under our present-day understanding of what actually constitutes "our heritage." The more strongly and deeply idealistic a man is, the more genuinely and sincerely he holds his honest beliefs, the more valiantly he will defend these "truths" that are, to him, self-evident.

Social inventions are most desperately needed today—and are hardest of all to make, because each man, within himself, has limited his own creative thinking. By failing to find the fundamental core of his ideals, he may sacrifice everything in a pointless defense of a nonessential.

Fifty years ago, the engineering student was considered something of a second-class citizen of the college campus; only the Liberal Arts student was considered a true student. A social invention was making its way, however. Where major corporations and businesses were uniformly directed by lawyers and Liberal Arts students only one generation ago—today the technical man is taking a bigger and bigger part in executive control.

Educational methods, more than any other single factor, will determine what our world is like in another half century. Of all possible forms of education, it seems to me that the most critical is education to understand, use, and evaluate creative thinking.

It is my feeling that studies of creative thinking itself—such work as Professor Arnold and his co-workers at M.I.T. have started—are basic to understanding our Research Age civilization. Where such work as Newton did was necessary to understanding the physical world, studies of creative thinking are necessarily more fundamental; understanding gravity did not necessarily lead to understanding creative thinking. But if ever Mankind learned to understand creative thinking, that necessarily implies ability to generate an understanding of all physical forces.

No full solution to the problem of understanding creative thinking yet exists—but the M.I.T. group has launched a solid, conscious and directed attack on that problem. It's an engineering attack—"A theory that works may not be true, but it's useful until a better theory can be developed."

THE EDITOR.

Science fiction in the classroom? What! You're designing for nonhumans on far distant planets? Aren't there enough unfulfilled human needs that you could design for and thereby better use your time? These are some of the typical questions that are asked when people first hear of the Arcturus Project used in the Product Design Course at M.I.T. After explaining the project and the course, however, these questions usually change to exclamations such as: "What an idea, I wish

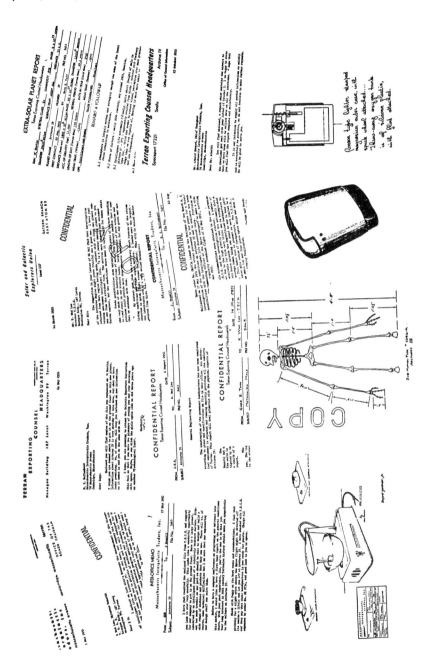

I could have taken a course like that!"

This course is relatively new at M.I.T., just three years old, and is part of a rapidly expanding program in creative engineering. The program started with one course elected by seniors and by next fall it will consist of a sequence of courses starting in the sophomore year. This rapid expansion is the result of the encouraging evidence presented by the initial experimental course. It *is* possible to train students to think more creatively; one *can* develop his imagination. The most encouraging aspect of the experiment is that this is as equally true for many who originally thought that they had little talent for design as it is for those who had previously exhibited a high order of imagination. The students claim that they leave the course with a new perspective with which to face a broad variety of problems.

Before describing the course in some detail it would be wise to define some of the terms that will be used repeatedly. Science and engineering like most other fields of endeavor use two main thinking processes, analytical and creative. They are quite different and should be carefully defined. There is a third important process, the judicial, that contains aspects of the two above and is used in conjunction with them to help insure meaningful results.

There are three ways to distinguish whether a problem is analytical or creative: first, the statement of the problem; second, the approach used in its solution; and third, the results obtained. An analytical problem is stated in quite definite terms—determine the deflection at the center of a given beam under uniform load conditions. The creative problem expresses a need—it is desirable at times to have the surfaces of sliced bread browned, heated and dehydrated. The approach used in the first problem is as definite as its statement. Knowing all the physical properties and dimensions of the beam, its span, and constraints and load per unit length, the straightforward application of $\dfrac{d^2y}{dx^2} = \dfrac{M}{EI}$ will yield the desired result. A second type of approach is frequently used in the analytical problem, that of building a model of the prototype, or using the prototype itself, loading it per specifications and then measuring the desired result.

The approaches to the solution of a creative problem may be without limit. Everyone knows that the use of the radiant energy of an electrical resistance element will solve the type problem listed above, but this is by no means the only way to solve that problem. It is possible that some chemical mixed with the butter—or any other spread—might do the job as well or even better. Maybe high-frequency heating, or slicing the bread with heated wires would be equally as effective. Changes in the structure or composition of the bread itself should not be overlooked in solving the expressed need.

Looking at the results obtained is probably the easiest way to distinguish between an anlytical and creative problem. Taking into account the state of the art of any particular time, there is only *one* right answer to an analytical problem. The solutions to a creative problem, on the other hand, may form a complete spectrum, depending on the thoroughness with which it has been investigated. It is impossible to say that any one answer is *the* right answer, continued investigation may lead to a better one.

To summarize then, the analytical problem is very specific in its statement; two approaches are usually employed in its solution, a process of logical reasoning or one of empirical testing; and, within the existing state of the art, there is only *one* right solution. Ninety per cent or more of all the courses taught in our public schools and colleges deal with problems of this type. The creative problem, in contrast, is stated in very nebulous, very general terms. It implies or expresses a need in such a way that almost an infinite number of specific approaches may be formulated and carried out in search of a solution. The results obtained run the gamut from good to poor and there is always room for new approaches to better solutions. Very few courses attempt to handle problems of this type although the need for creative thinkers is as

great, if not greater, than for those of the analytical type. The statement of this need implies a creative problem from the very start and the solution described below, by definition, is not the one, right solution. The results obtained indicate that it is a good one, but the search goes on for a better one.

The aim, then, of the M.I.T. Creative Engineering program is to provide an ever increasing number of young men trained, not only in the basic concepts of science and engineering but also in the use of their creative imaginations, to help solve the ever increasing problems, both in complexity and in numbers, that continue to face the nation and the world. Design courses provide an ideal vehicle for this kind of training, but by no means should this training be restricted to this field.

The Product Design course is conducted in an informal seminar fashion. Three two-hour seminars are held each week. These are devoted to discussions, demonstrations and laboratory work, so that the student will learn first, how does one think creatively and what is the creative process; second, what tools does the creative engineer work with and what factors should he take into consideration in the solution of his problems; and third, that through constant practice he will become more proficient in exercising his imagination and will gain confidence in his ability to solve difficult, challenging problems.

It is not within the scope of this paper to discuss in detail the creative process and how it works. For those interested, a bibliography of recent papers and books on the subject is included at the end. A specific example, however, of how nonanalytical factors influence design will be given. Take for example the influence of semantics on the creative process. The students had been assigned, as one of their major design projects, a case study on a "Dual Sander." The case described in some detail some of the various types of sanding machines on the market, pointing out their good and bad features. Two types of machines were singled out for specific analysis, the rotary disk type and the vibrating plate type. Sufficient technical data and a list of desirable design specifications were included so that the students could confidently design either type of machine.

The case then pointed out the desirability of combining the two types of motion into one "all-purpose" machine. This machine would provide fast, rough sanding—disk type—and fine, finish sanding—vibrating type. Layouts for three possible solutions to this problem were included for the students' guidance or criticism.

After the students had had an opportunity to read the case, one full seminar session was devoted to discussing the case in particular and

finishing methods in general. At the beginning of the next seminar session the students were asked to write in their own words in as general terms as possible the aim of this project. They were also asked to list in outline form a method of attack for solving their listed goal.

The majority of students put down quite specific aims and very definite *modus operandi*, this in spite of previous seminars on semantics and the very definite instructions given them before they were asked to write. A typical aim listed by this group was, "The aim of this project is to design a dual purpose *sanding* machine, to provide rough and fine *sanding*." A prosaic, standard approach was listed as the method of attack. These students without realizing it were greatly limiting themselves at the beginning of the creative process. This was in part due to the pre-conditioning effect that the case study had on their thinking.

A small group of students, however, were able to ignore the original statement of the problem and set up for themselves a new goal that gave almost unlimited scope to the problem. "The aim of this project is to design a multipurpose smoothing machine or process. Smoothing may be accomplished by either adding or removing material." A statement of this type naturally leads to a very general approach and the search for a solution would enter every technical field,

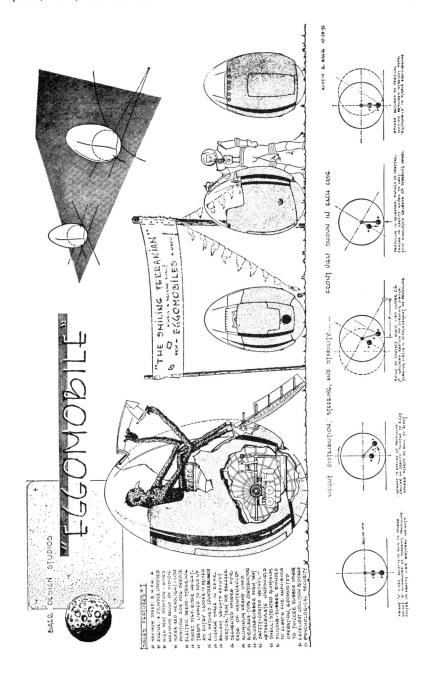

electrical, chemical as well as mechanical.

The remainder of this seminar was devoted to exploring the possibilities opened up by the more general type of goal. This was accomplished by everyone first listing all possible ways by which any material could be smoothed by adding or subtracting material. The listing was done without any critical evaluation of the method as to its practicality or even feasibility or to its economics. As the lists were individually read off new ideas were added until it was felt that the various fields had been fairly well exhausted. Then and only then was critical analysis applied to the many suggestions.

Of course, many of the proposed schemes had to be discarded because of their impractical nature or the possible high costs associated with them, but many of the suggestions were developed to a point where they looked as though they would be very profitable avenues for further research. A few had rather limited fields of application and were far removed from sanding machines. In many cases it was discovered that the proposed methods had already been incorporated in various machines and processes, e.g. smoothing by adding material, metal spray guns, and smoothing by mechanical compression, calendering of paper.

It was decided that actually Emerson's advice "to build a better mouse trap" would not have led to the "path-beating-act" unless the inventor had restated his problem in more general terms. Emerson's statement would have insured a trap being built but would have precluded the possibilities of electrocuting, poisoning, drowning or even frightening the mice to death.

The case that has had the widest publicity and engendered the most discussion is the Arcturus IV Project. It was designed, in part, to free the student of all preconceived notions about man-machine relationships and to strengthen the influence of environment on design. There are many other reasons for introducing a case of this type and they will all be discussed after describing the case.

Arcturus IV is the fourth planet out from the sun α Bootis (Arcturus), thirty-three light-years from our solar system. It was first contracted by a member of the Solar and Galactic Explorers' Union on January 22, 2951. It is a large planet, 12×10^6 meters in diameter, having a mass of 60×10^{27} grams and the acceleration of gravity at the surface is eleven thousand centimeters per second squared. It is a distance of 1800×10^6 miles from α Bootis and its siderial period is 49.4 Earth-years. The length of day is one hundred fifty-nine hours; the atmosphere is largely methane; and the mean temperatures range from $-50°$ C in the summer to $-110°$ C in the winter.

All the information about the planet and its inhabitants is obtained from the files of the Massachusetts Inter-galactic Traders, Inc. and each student receives a copy of this file. M.I.T. Inc. is engaged in the manu-facture and distribution of products for extrasolar consumption. (For the students' benefit the products must be manufactured using Twentieth Century technology and materials.) This company and all others like it operate under the rules and regula-tions of the Terran Exporting Counsel Headquarters, a government agency. T.E.C.H. sets up a branch office on all planets with which Terra is doing business and its divisions such as the General Engineering Division, Physio-logical and Psychological Division, and the Design, Production and Mar-keting Division carry out detailed investigations and write and publish reports for all who might be inter-ested. These are included in the files.

In drawing up this case study every effort was made to make everything as realistic and consistent as possible. So far no glaring errors have been dis-covered. All information in the file is on specially prepared stationery and report forms, stamped and handled in the best businesslike manner. The only thing that is lacking is reliable market reports on the sales and ac-ceptance of the products designed.

The race of people—subhuman, of course—that inhabit the contacted portion of Arcturus IV are called Methanians. A good description of them is contained in a report from J. S. Wick, Director of the Physio-logical and Psychological Bureau of T.E.C.H. "Strangely enough the Me-thanian metabolic process is similar to Terran plant life. Carbon is obtained from the Methane atmosphere and oxygen from the plant and animal life eaten as food. There is no liquid water anywhere on the planet and due to the very cold temperature, little in the atmosphere. The water that is present is in the solid state resulting in a foggy condition both winter and summer. Ammonia is the Arcturian substitute for water.

"The Methanians weigh very little compared to us. One of the largest we met was weighed on a Terranian spring scale at one hundred eighty-seven pounds. (They are relatively strong, however, being able to lift twice their own weight.) Their bones are hollow and apparently filled with hydrogen and helium. There is no question but these people have evolved from a race of birds, their appearance seems to indicate it, their history seems to prove it. Their long arms and clawlike hands—three-fingered—are vestiges of once great wings. The only anomaly is their single-toed feet like that of a horse. This adaption to ground living evolved very rapidly once the power of flight was lost.

"The young are born in eggs and the eggs are carried around in skin pockets or pouches similar to those of the now extinct Terranian Penguin until the egg hatches. Both male and female take turns in the hatching process. The young grow rapidly at first and are ready to take care of themselves in about twenty Terranian years. They seldom leave home, however, before physical maturity is reached, 49.4 Terranian years.

"The Arcturian normal body temperature is − 40° C and their pulse rate is five times per minute. As a result they are very slow-moving and they frequently walk using one or both arms as a cane or pair of crutches. Their normal walking pace is about one fourth mile per hour, but if pressed can go almost eight times as fast for very short periods. Even with HI-G units we don't travel much faster than they do. This slow pace does not seem to bother them since their whole system is geared to it. Their stimulus response time is about two seconds.

"Their auditory, vocal and visual range is extremely large. They can hear sounds with frequencies as low as 1/100 cycle/second up to 50,000 cycles/second. Their vocal range goes from 1/50 to 25,000 cycles/second and their visual range extends from the infrared up through the ultraviolet.

"As you might expect they are very stable emotionally, very slow to anger and with tremendous patience measured by our standards. They have a limited amount of telepathic ability but seem to use this form of communication only under duress. In the ESP tests we thought we had discovered a race with exceptional talent but later found out that their high, almost perfect, scoring was due to the X ray-like vision of the third eye."

The reports and letters in the file try to cover briefly, of course, most of the important phases of the life and culture of the Methanians and the physical features of the planet. As the students design, however, new information is frequently needed and it is part of the student's job to provide this information consistent with that already given. The first design project was limited to products of a household or personal use classification. One of the students wanted to design a clock for the Methanians and consequently was forced to devise a logical subdivision of the Methanian day and a numbering system for them. A portion of his report follows:

"The number system is based upon six (6), as would be suspected upon considering Methanian three-digit hands. The number system definitely evolved from finger counting. Thus 1, 2, and 3 are ı, v, and ♥. An alternate symbol for three was the closed fist, which gradually deteriorated into a small circle. Thus, four would be one finger and one fist or 01. This gradually became ᴏ. Similarly, five is ᴏᴠ and six would be two fists. This ultimately became two circles, one on top of the other, or 8. This is, of course, our figure "eight" exactly. The idea of building up larger numbers by ar-

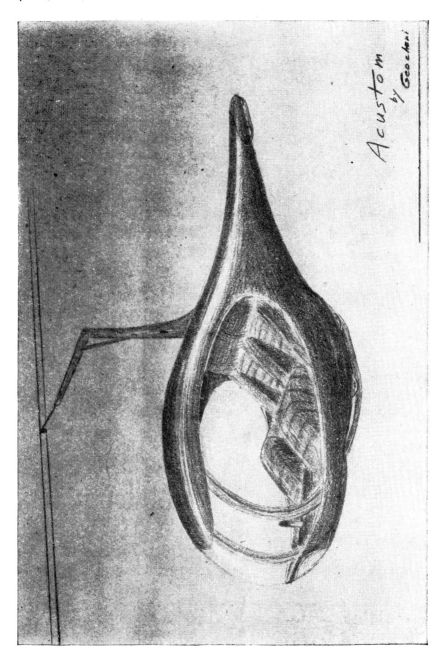

ranging symbols in sequence, and allowing the position of the symbol to indicate its value—as we do in Arabic notation — was introduced about eighty-five Methanian years ago and the symbol 8 became the zero. The complete history of the development of the number system is interesting, but only the final result is given here:

GINT and is equivalent to 2.378 ± .001 volts. Electrical transmission over great distances is accomplished with an emf of 1296 GINTS = 3080 volts. Individual house voltage is stepped down to 1/36 of this, 85.6 volts. This standard emf of thirty-six GINTS seems to be available in almost every Methanian structure which uses electricity."

	1	I	7	II	13	VI	19	¥I	25	αI	31	αVI	
	2	V	8	IV	14	W	20	¥V	26	αV	32	αVV	
	3	¥	9	I¥	15	VV	21	¥¥	27	αV	33	αVV	
0=8	4	α	10	Iα	16	Vα	22	¥α	28	αα	34	αVα	etc.
	5	αV	11	IαV	17	VαV	23	¥αV	29	ααV	35	αVαV	
	6	I8	12	V8	18	¥8	24	α8	30	αV8	36	I88	

The reports in the file indicated that the Methanians used electricity generated from atomic power plants, but no details of the system were given. This same student, in order to power his clock with electricity, had to fill in the missing details. "Alternating current is generated. The frequency of the current is 1296 cycles per NAHLO —the shortest subdivision of the Methanian day. Note that $1296 = 6^4 = 18888$ in Methanian notation. This is analogous to 10000 in Terranian notation. Since one NAHLO = 7.37 minutes = 442.5 seconds, than $1296/N = 1296/442.5 = 2.926$ cps which is about 1/20 of standard Terran frequency of 60 cps. The Methanian electrical science is based upon the concepts of emf, current and resistance. Their unit of emf is called the

The description of the electrical system above combined with the information previously given about the Methanians gives rise to one possible inconsistency which is left to the reader to argue out for himself. Considering the very slow stimulus-response time, the wide range of auditory and visual reception and the very slow electrical frequency, would the Methanians be bothered by flicker from their electric lights? Another similar question is, at what frequency should motion pictures be projected?

Some of the other designs carried out by the first group of students subjected to this problem were chairs and tables, two different telephone designs, kitchen food-mixers, combination egg-incubator and baby-stroller, a stereo slide viewer and a complicated "lawn-

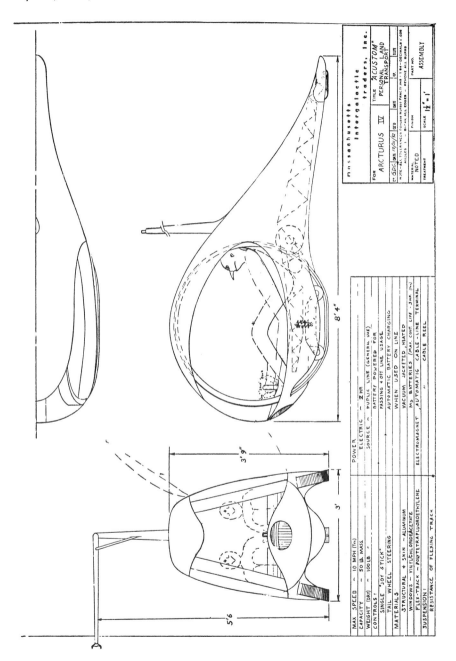

Massachusetts Intergalactic traders, Inc.

TITLE "ACUSTOM" PERSONAL LAND TRANSPORT

FOR ARCTURUS IV

DR SDG DR 10/21/52 CKD

NOTE—ALL TOLERANCES (UNLESS NOTED) FRACTIONS ± 1/64—DECIMALS ± .008
BUT ± ALL EDGES — REMOVE ALL BURRS

MATERIAL NOTED

FINISH

TREATMENT

SCALE 1 1/2" = 1'

PART NO.

ASSEMBLY

8' 4"

3' 9"

3'

5' 6"

POWER

ELECTRIC ~ 2 HR

SOURCE ~ PUBLIC LINE (GENERAL USE)
BATTERY POWERED FOR
PASSING 4 OFF LINE USAGE
AUTOMATIC BATTERY CHARGING
WHEN USED ON LINE

MATERIALS

STRUCTURAL + SKIN — ALUMINUM
WINDOWS — VINYLCHLORIDEACETATE
FLEX-TRACK — POLYTETRAFLUOROETHYLENE

MAX SPEED ~ 10 MPH (Tr.)
CAPACITY ~ 50 LB. MASS
WEIGHT (DRY) ~ 100 LB. "

CONTROLS:
SINGLE "JOY STICK"
TAIL WHEEL STEERING

VACUUM JACKETED, HEATED
M4 BATTERIES (MAX. CONT. LINE ~ 3 HR. (Tr.))

ELECTROMAGNET ,AUTOMATIC CABLE-LINE TERMINAL
" CABLE REEL

SUSPENSION:
RESISTANCE OF FLEXING TRACK

conditioner" for the upside down Methanian vegetation. In all cases the designs had to work and had to meet the exacting conditions of the Arcturus IV environment as well as being adapted to the Methanian's physical and psychological limitations. They had to be built with present-day technology and with materials now available on Earth. Weight limitation was hard to meet; temperature limitation caused the most trouble in getting reliable information on materials.

The second group of students to design for Arcturus were all asked to design means of powered transportation for the Methanians. The idea of introducing "automobiles" to a primitive culture that had never used anything but foot-power and domestic animals caused a great deal of discussion in the seminars. Would it be possible and desirable to introduce a highly perfected machine or should the introduction follow the history of the development of the automobile on Earth? An expert on primitive cultures was brought in to lead one of the seminar sessions and a furious battle was fought. Neither side won a clear victory so that designs following the two approaches were submitted. The Eggomobile pictured on the cover and in the accompanying plate was typical of the "conservative" approach. Due to its shape and resulting stability problems it is limited to very low speeds and changes in momentum.

But again, its egg shape would be psychologically desirable and give the Methanians a sense of complete security, a very important factor in introducing these "demons" of the road.

The little Acustom Coupe pictured on page 19 on the other hand, is capable of very high speeds and accelerations—see limitations below—is very efficient in design and is typical of the "damn history" approach to design. The problem of roads is a difficult one that must be faced by the designer when he attempts to introduce a powered vehicle into a society that is used to going about slowly on foot on narrow paths. The large spherical drive unit of the Eggomobile and the flexible treads of the Acustom would make them adaptable to most any terrain. The Acustom is limited by its electric motor and trolley pickup to previously laid out paths and brings up the question of what one does when he meets or wishes to pass another vehicle. Most of the vehicles were powered by internal combustion engines or gas turbines, the fuel being hydrogen peroxide.

One of the students felt that due to the Methanians egg-birth they would hold the egg and all similarly shaped objects in the deepest reverence. It would, therefore, be bordering on the sacrilegious to use the wheel for such a lowly job as transportation. As a consequence, he designed a machine that propelled itself by walking. It was a comparatively simple design with an ingenious system for turning. The ride was described as being similar to that obtained with a Terranian Camel although not quite as comfortable.

The major limitation in this car design was the very slow stimulus-response time of the Methanians. Without the use of automatic controls —that was too much of a new concept to introduce at this time—how fast should they travel and still be able to avoid hitting stationary or moving objects? In starting this discussion it was argued for some time whether or not the Methanians could even stand upright and walk. Considering the slow s-r time, the high acceleration of gravity and their high center of gravity, the poor Methanian might be flat on his face before he knew he was falling. It was finally decided, however, that the Methanian would have developed some anticipatory sense similar to that developed by the human child when learning to walk. The use of his long arms in walking, of course, increases his stability. Is it possible to apply similar reasoning to driving a car? The answer was yes.

The human being in learning to drive a car is consciously dependent on stimulus-response mechanisms to keep him going in a straight line and frequently overcorrects the detected errors. With practice, however, the subconscious soon takes over and errors are corrected almost before they

are large enough to be detected. The amazing computing capacity of the brain is able to solve in a fraction of a second the many simultaneous equations that must be solved in order to pass safely through an intersection loaded with pedestrian and vehicular traffic. The equations involved might take days of conscious effort to solve. It was decided, therefore, that the Methanian could develop in a similar fashion over a period of time.

There was some question as to whether the Methanian brain could ever work as fast as the human brain because of the low metabolic rate and s-r time. It was arbitrarily decided that the maximum speed of all vehicles—subject to subsequent testing —be limited to fifteen miles per hour. It is very likely that this high speed would not be reached until a number of years of adjustment had passed by.

The reader can very likely imagine many other points that should be considered in designing for the Methanians but he can be assured that the chances are very good that the designers of the Massachusetts Intergalactic Traders, Inc. have given them due consideration. Do you think that the average, present day Terranian designer gives as much thought to human limitations?

The Arcturus IV project accounts for about one-fourth of the student's time in this course. The other three-fourths of the time is devoted to more prosaic, earthly designs. Yet the three weeks or so spent out in space are richly rewarding and have a distinct carry-over value and a profound influence on the remainder of the course. The case was first set up because the answer to the question posed in the paragraph above seemed to be no. It was hoped that a dramatization of this type would forcibly bring home to the student the importance of the man-machine relationship and the influence of environment.

Some of the seminars held while the Arcturus case was in progress were devoted to examining some of the results of the Applied Psychologists of the Tufts College group and of the Special Devices Center for the Navy on Long Island. The students were amazed to see how much had already been done in the field of human-engineering or bio-mechanics, as it is sometimes called. They also realized that there is a great deal more to be done.

It is very difficult to accurately measure the influence of this one case on the students' subsequent thinking in the field of human engineering, but a qualitative measure can be obtained by sitting in on any one of the later design seminars, be it on Sanding Machines, Rug Shampooers or Turbo Cars, and comparing it with any other typical design group, in or out of schools. The enthusiasm for detail and the relentless search for all the factors, nonanalytical as well as analytical,

that might influence a design is a very encouraging sight.

There have been many other beneficial results obtained with this first experiment with science fiction in the classroom. First of all, it provides a very stimulating jolt to the imagination, a jolt which some students probably couldn't survive. The more imaginative a boy is the quicker he adjusts himself to this new situation. The big adjustment demanded by the Arcturus case makes the subsequent adjustments relatively easy. He has to stretch his imagination to such a limit that it doesn't quickly shrink back to its former inconspicuous self.

In the second place, since it is almost impossible to prove or disprove some of the controversial issues that are raised by the Arcturus case, a student who conscientiously bases his design on principles which he thinks are logical and sound gains a confidence in his ability to design rationally and creatively that the most vicious design jury cannot destroy. This confidence in one's ability is one of the prime prerequisites for all good designers. If one doesn't have it or can't develop it, he had better look for something else to do. The weight of evidence that could be brought to bear by a design jury against a mistaken design principle used in an Earth-consumed product could materially affect the quality of the designer's subsequent work by shaking his confidence in himself. In the

Arcturus case the student designer can always rationalize that he is as much entitled to his opinion as the jury is to theirs and everyone lives happily ever after.

And lastly, a great many of the students with imagination are already science-fiction fans or else take to it very readily. The result is that the first case he works on is fun and not work; he learns while he enjoys himself. There may be some theory that education must be solemn and serious but the Creative Engineering Group at M.I.T. do not subscribe to it. The results of the informal seminars and lab sessions indicate that it would be desirable to hold all classes in a similar fashion. The Arcturus case is an excellent ice-breaker and strangers at the beginning of the course are good friends three weeks later.

It was indicated at the beginning of this paper that the program in Creative Engineering is expanding rapidly. This is due in part to the encouraging interest shown by all industries aware of the work that is being done. A number of grants have been received to be used for the preparation of new case material and other research and in one instance a large corporation went to considerable trouble and expense in the preparation of a very complete case history for one of the projects. The future of the program is limited only by the imagination of those participating in it, and this includes students as well as instructors. The course is designed for them, as every course should be, and they are encouraged to enter into its formulation which they freely do. The course then becomes a case study in Creative Engineering.

BIBLIOGRAPHY

Chapanis, Garner and Morgan—"Applied Experimental Psychology"; John Wiley and Sons, New York, 1949.

Easton, W. H.—"Creative Thinking and How to Develop It"; *Mechanical Engineering*, August, 1946.

Flesch, R.—"The Art of Clear Thinking"; Harper & Brothers, New York, 1951.

Killeffer, D. H.—"The Genius of Industrial Research"; Reinhold Publishing Corporation, New York, 1948.

Nine Authors—"Creative Engineering"; A.S.M.E., 1944.

Osborn, Alex. F.—"Your Creative Power"; Charles Scribner's Sons, New York, 1950.

Tufts College—"Handbook of Human Engineering Data for Design Engineers"; 1949.

Weinland, C. E.—"Creative Thought in Scientific Research"; *Scientific Monthly*, December, 1952.

THE END

★ Appendix ★

Recommendations for Further Reading

★The following lists have been provided by the indicated authors who have written chapters found elsewhere in this book. The lists have been edited such that after a first appearance of a title, subsequent lists do not repeat it. A few have been annotated by the editor.

Some supplementary comments might be helpful. Since chemistry is less well-represented in the science fiction genre than its cousins biology, physics, and the social sciences, these listings have invoked a very broad umbrella. For example, subatomic structures—a major domain of physics—may be considered to overlap chemistry. Similarly, aspects of biological mutation may legitimately be included in the chemistry domain. This appendix might be better described as one of good *science* fiction that includes as much chemistry-related material as possible.

Procurement of individual items may present serious challenges. Many of the novels, reflecting their generally superior quality, are reprinted regularly and thus are likely to be in print. Moreover, they can be obtained readily from many libraries or via interlibrary loan. The shorter works, particularly when they have not been chosen to provide the name of the collection, can present a more serious acquisition problem. To complicate matters further, the same story may appear in multiple collections and not always under the same name! Where a story received some major award (such as the Hugo or the Nebula) the annual Awards volumes provide a relatively ready access. Perhaps your best bet is to visit a fair-sized used paperback shop— these can be entertaining, the concentration of pertinent material can be substantial, and you might just luck into a knowledgeable proprietor.

Still another resource for the less resourceful: Almost all of the current science fiction magazines contain a classified ad section. Many of these ads offer a catalog, often extensive, of available SF books, hardcover and paperback, for free or a token amount. In addition, they solicit your "want lists".

List A: Provided by Connie Willis

Part 1. Chemistry-Related Stories

Aldiss, Brian. *The Eighty-Minute Hour*.

Anderson, Poul. "Death Wish". In *The Microverse*, Ed. Byron Preiss. New York: Bantam Books, 1989. Molecules.

Anderson, Poul. "Elementary Mistake". In *Analog 7*, Ed. John W. Campbell. Garden City, NY: Doubleday and Co., 1969. Strange new chemicals on a new planet.

Anvil, Christopher. "Not in the Literature". In *Analog 3*, Ed. John W. Campbell. New York: Doubleday and Co., 1963. Chemistry vs. physics, how the sciences are linked.

Asimov, Isaac. "As Chemist to Chemist". In *Isaac Asimov's Worlds of Science Fiction*, Ed. George Scithers. Using atomic numbers and the periodic chart to solve a puzzle.

Asimov, Isaac. "Take a Match". In *Buy Jupiter and Other Stories* by Isaac Asimov. Garden City, NY: Doubleday and Co., 1975. Chemical reactions.

Asimov, Isaac. "The Micropsychiatric Properties of Thiotimoline". In *Astounding Magazine*, December 1953. Second paper on endochronic filtration—trying to fool the thiotimoline—used people with split personalities; some thiotimoline dissolved, some didn't.

Asimov, Isaac. "Thiotimoline and the Space Age". In *Year's Best SF: 6th Annual Edition*. New York: Simon and Schuster, 1961. Dissolves one whole day before water added; used for weather forecasting and predicting horse races.

Asimov, Isaac. "Thiotimoline to the Stars". In *Buy Jupiter and Other Stories* by Isaac Asimov. Its application for space travel.

Ballard, J. G. "The Voices of Time". In *The Best of J. G. Ballard*. London: Futura Publications Ltd., 1977. Entropy.

Benford, Greg. *Timescape*. Tachyons and high-energy ions.

Blish, James. "Beep". In *Galactic Cluster*. New York: Faber and Faber, 1963. Positrons.

Blish, James. "Nor Iron Bars". In *Best Science Fiction Stories of James Blish*. London: Faber, 1965. Spacecraft reduced in size so that it can travel in the solar system of the atom.

Brown, Fredric. "Placet is a Crazy Place". In *The Astounding-Analog Reader*, Ed. Harry Harrison and Brian Aldiss, 1972. Matter and antimatter.

Brunner, John. "Report on the Nature of the Lunar Surface". In *First Flights to the Moon*, Ed. Hal Clement. Garden City, NY: Doubleday and Co., 1970. Chemical composition of the moon's surface.

Caravan, T. P. "In Happy Solution". In *Other Worlds*. To get an A in thermodynamics, hero needs to find a lock of hair from a bald professor and a container for a universal solvent.

Chandler, A. Bertram. "Critical Angle". In *First Flights to the Moon*, Ed. Hal Clement. Kinetic energy.

Chapman, Steve. "Testing... One, Two, Three, Four". In *Analog 8*, Ed. John W. Campbell. Garden City, NY: Doubleday and Co., 1971. Chemical warfare.

Clarke, Arthur C. "The Fires Within". In *Worlds of Tomorrow*, Ed. August Derleth. New York: Pellegrini and Cudahy, 1953.

Cooper, Ralph S. "The Neutrino Bomb". In *Great Science Fiction by Scientists*, Ed. Groff Conklin, 1961. Neutrinos.

Crichton, Michael. *The Andromeda Strain*. Chemistry and biochemistry are used to help discover the nature of the virus; mass spectrometer, carbon compounds, pHs.

Harrison, M. John. "Running Down". In *The Machine in Shaft 10 and Other Stories*. London: Panther Books Ltd., 1975. Entropy.

Hoyle, Fred. "Element 79". In *Element 79* by Fred Hoyle. New York: New American Library, 1967. Mock history of the discovery of a new element and its consequences.

Large, E. C. "Sugar in the Air". 1937. Artificial synthesis.

Lee, William. "Junior Achievement". In *Analog 2*, Ed. John W. Campbell. Garden City, NY: Doubleday and Co., 1962. Chemistry and kids.

Lem, Stanislaw. *Solaris*. (This is the story that most effectively provides the quality of alien-ness—Editor.)

Malzberg, Barry N. "Varieties of Technological Experience". In *Microcosmic Tales,* Ed. Isaac Asimov. New York: Taplinger Publishing Co., 1980. Universal solvent.

Myers, Howard L. "Out, Wit!". In *Analog 9,* Ed. Ben Bova. 1973. Alchemy—turning things to gold.

Niven, Larry. *The Smoke Ring* and its sequel *The Integral Trees.* Sulfur and oxygen atoms, properties of gases, surface tension.

Niven, Larry. "Unfinished Story No. 1". In *All the Myriad Ways.* Maxwell's Demon.

Piper, H. Beam. "Omnilingual". In *Great SF Stories about Mars,* Ed. T. E. Dikty. New York: Frederick Fell, 1966. A science-as-plot story involving a long-dead Martian civilization and the periodic table as a Rosetta stone. (Strongly recommended—Editor.)

Poe, Edgar Allan. "Von Kempelen and His Discovery" (1849). In *The Science Fiction of Edgar Allan Poe,* Ed. Harold Beaver. 1976. Consequences of alchemy.

Pohl, Frederik. *The Coming of The Quantum Cats.*

Shaw, Bob. *A Wreath of Stars.*

Silverberg, Robert. "Chip Runner". In *The Microverse,* Ed. Byron Preiss. New York: Bantam Books, 1989. Electrons.

Thomas, Theodore L., and Kate Wilhelm. *The Year of the Cloud.* The chemical properties of water.

Vance, Jack. "The Potters of Firsk". In *Time Probe: The Sciences in Science Fiction,* Ed. Arthur C. Clarke. New York: Delacorte Press, 1966. About the many uses of uranium; mystery in which clue is the bright yellow color of uranium salts.

Vonnegut, Kurt, Jr. *Cat's Cradle.* Chemical properties of water.

Wells, H. G. "The Diamond Maker." In *The Complete Short Stories of H. G. Wells.* London: Ernest Benn Ltd., 1966. Turning carbon into diamonds.

Wells, H. G. *The Food of the Gods.* Chemical principle of growth.

Wilson, Robert Anton. *Schrödinger's Cat.* Quantum theory.

Part 2: Selected Bibliography of Science-Related Science Fiction

Short Stories

Anderson, Poul. "Plato's Cave". In *Foundation's Friends,* Ed. Martin H. Greenberg. New York: Tor Books, 1989. A story using Asimov's three laws of robotics and the problems they give rise to.

Asimov, Isaac. "The Bicentennial Man". In *The Arbor House Treasury of Modern Science Fiction*, Eds. Robert Silverberg and Martin Greenberg. Garden City, NY: Doubleday and Co., 1970. One of Asimov's best robot stories, in which he deals with the three laws and the nature of what is human.

Asimov, Isaac. "Nightfall". In *The Science Fiction Hall of Fame: Volume 1*, Ed. Robert Silverberg. Garden City, NY: Doubleday and Co., 1970. A science-as-subject classic that deals with a planet that has several suns and sees the night sky only once in a thousand years.

Blish, James. "Surface Tension". In *The Arbor House Treasury of Modern Science Fiction*, Eds. Robert Silverberg and Martin Greenberg. Garden City, NY: Doubleday and Co., 1970. Humans downloaded into microscopic creatures who live in tidal pools and battle rotifers. A classic that uses science both as subject matter and as metaphor.

Brown, Fredric. "The Waveries". In *Invaders of Earth*, Ed. Groff Conklin. New York: Vanguard Press, 1952. Aliens who eat electromagnetic radiation descend on earth, attracted by our radio and TV transmissions; chronicles the history of radio and TV.

Brown, Fredric. "The Yehudi Principle". In *Angels and Spaceships*, New York: E. P. Dutton, 1954. A time paradox story in which the story itself is part of a time loop with no beginning and no end.

Bryant, Ed. "Particle Theory". In *Nebula Winners 13*, Ed. Samuel Delaney. New York: Harper and Row, 1980. A science-as-metaphor story which compares the phenomenon of supernovae with human cancer.

Clement, Hal. "Blot". In *Foundation's Friends*, Ed. Martin H. Greenberg. New York: Tor Books, 1989. A robot story that gives a variant on Asimov's three laws of robotics.

Dickson, Gordon. "Computers Don't Argue". In *Analog's Golden Anniversary Anthology*, Ed. Stanley Schmidt. New York: Davis Publications, 1980. A very funny technology-run-amok story about the problems that can result from dealing with too-literal computers.

Dieppe, Carol, and Lee Wallingford. "Special Delivery". In *Isaac Asimov's Science Fiction Magazine*, August 1989. A science-as-plot-device story about a secret code hidden in the DNA sequence.

Effinger, George Alec. "Schrödinger's Kitten". In *The Year's Best Science Fiction: Sixth Annual Collection*, Ed. Gardner Dozois. New York: St Martin's Press, 1989. A science-as-metaphor story about relativity and alternative realities based on Schrödinger's cat-thought experiment.

Fitzpatrick, R. C. "The Circuit Riders". In *Analog 2*, Ed. John W. Campbell. Garden City, NY: Doubleday and Co., 1964. A future-technology story that deals with a machine that detects emotions and its possible applications.

Godwin, Tom. "The Cold Equations". In *The Science Fiction Hall of Fame, Volume 1*, Ed. Robert Silverberg. Garden City, NY: Doubleday and Co., 1970. A science fiction classic that deals with the sometimes-tragic human dilemmas that result from physics's inexorable laws. (There is a current movie of the same name, but it has seriously changed the original story—Editor.)

Harness, Charles. "Child by Chronos". In *The Best from Fantasy and Science Fiction, Third Series*, Eds. Anthony Boucher and J. Francis McComas. 1952. A very cleverly worked-out time-paradox story about a person who travels into the future, memorizes the daily stock market reports, and then returns to the past to make investments.

Harrison, Harry. "The Fourth Law of Robotics". In *Foundation's Friends*, Ed. Martin H. Greenberg. New York: Tor Books, 1989. An examination of the logical flaws in the three laws of robotics.

Heinlein, Robert A. "All You Zombies". In *The Arbor House Treasury of Modern Science Fiction*, Eds. Robert Silverberg and Martin Greenberg. New York: Arbor House, 1980. The ultimate time-paradox in which a man begets himself.

Heinlein, Robert A. "By His Bootstraps". In *The Menace from Earth*. New York: New American Library, 1959. A classic time-paradox story with one person playing all the parts.

Heinlein, Robert A. "Waldo". In *Waldo and Magic, Inc.* New York: Pyramid Books, 1950. A science fiction story that not only predicted the future but invented it. Because of this story, "Waldos" became the term for the remote-action tools used in nuclear and micromanufacturing.

Latham, Philip. "The Xi Effect". In *Best Science Fiction Stories*, Ed. Edmund Crispin. London: Faber and Faber, 1962. A science-as-subject story in which the universe begins shrinking but the length of the electromagnetic spectrum stays the same.

Niven, Larry. "Man of Steel, Woman of Kleenex". In *N-Space*. New York: Tor Books, 1990. A very funny story that examines Superman's powers in terms of scientific laws.

Reynolds, Mack. "Compounded Interest". In *SF 57: The Year's Greatest SF and Fantasy*. New York: Gnome Press, 1957. A time-travel story in which a man takes money back into the past to invest so the interest that it accumulates can pay for the time machine that will take him into the past to invest the money.

Rocklynne, Ross. "The Men and the Mirror". In *Before the Golden Age*, Ed. Isaac Asimov. Garden City, NY: Doubleday and Co., 1974.

Shaw, Bob. "The Light of Other Days". In *The Arbor House Treasury of Modern Science Fiction*, Eds. Robert Silverberg and Martin Greenberg. New York: Arbor House, 1980. A science-as-subject classic that extrapolates the uses and human consequences of glass with a slowing index of refraction.

Vonnegut, Kurt, Jr. "Harrison Bergeron". In *The Best from Fantasy and Science Fiction, 11th Series*, Ed. Robert P. Mills. Garden City, NY: Doubleday and Co., 1962. An if-this-goes-on story that puts the idea of absolute human equality into a possible technological future.

Willis, Connie. "At the Rialto". In *The Year's Best Science Fiction, Seventh Annual Collection*, Ed. Gardner Dozois. New York: St. Martin's Press, 1990. A science-as-metaphor story about the peculiar effects of quantum theory on a physicists' convention in Hollywood.

Willis, Connie. "Dilemma". In *Foundation's Friends*, Ed. Martin H. Greenberg. New York: Tor Books, 1989. A robot story using Asimov's three laws that examines the dilemma created by two directly conflicting orders.

Willis, Connie. "Schwarzschild Radius". In *Nebula Awards 23*, Ed. Michael Bishop, New York: Harcourt Brace Jovanovich, 1989. A science-as-metaphor story about Karl Schwarzschild, the scientist who, while serving on the Russian front in World War I, extrapolated Einstein's theory to produce the concept of a black hole.

Willis, Connie. "The Sidon in the Mirror". In *The Year's Best Science Fiction, First Annual Collection*, Ed. Gardner Dozois. New York: Bluejay Books, 1984. A story that uses Harlow Shapley's theory of stellar evolution to set a story on the cooling crust of a red giant.

Zoline, Pamela. "The Heat Death of the Universe". In *The Mirror of Infinity: A Critics' Anthology of Science Fiction*, Ed. Robert Silverberg. 1970. A classic science-as-metaphor story comparing a housewife's daily routine and entropy.

Novels

Asimov, Isaac. *Fantastic Voyage*. Boston: Houghton Mifflin, 1966. A science adventure in which a spaceship is shrunk to microscopic size and injected into a human body.

Asimov, Isaac. *I, Robot*. Garden City, NY: Doubleday, 1963. The three laws of robotics and the complications that result from their application.

Clarke, Arthur C. *Rendezvous with Rama*. New York: Harcourt, Brace, Jovanovich, 1973. The classic science-as-subject novel of scientists exploring and attempting to understand an abandoned alien spaceship. (There have since appeared three sequels: *Rama II, Garden of Rama,* and *Rama Revealed*—Editor.)

Crichton, Michael. *The Andromeda Strain*. New York: Random House, 1969. High-tech science-as-subject story of the scientific effort to isolate an alien virus.

Dick, Philip K. *Do Androids Dream of Electric Sheep?* (published as *Bladerunner*). New York: Ballantine Books, 1968. A classic dystopian novel of androids and artificial life forms. Science as subject and as metaphor.

Gernsback, Hugo G. *Ralph 124C41+*. London: Cherry Tree, 1952. The novel that started it all and started SF's reputation for predicting the future.

Heinlein, Robert A. *Have Space Suit, Will Travel*. New York: Scribner, 1958. A space adventure full of science: astronomical units, breathable H/O levels in spacesuits, calculating gravities, plus a keen love of and interest in science.

Heinlein, Robert A. *Time for the Stars*. New York: Ace Books, 1956. A space adventure dealing with relativistic effects of traveling near the speed of light.

Niven, Larry. *Ringworld*. New York: Ballantine Books, 1970. Science-as-subject space adventure of a team exploring an artificial ring constructed around a sun. (Two sequels, *Ringworld Engineers* and *The Ringworld Throne* have appeared—Editor.)

Stith, John. *Redshift Rendezvous*. New York: Berkeley Publications, 1990. A murder mystery set near the speed of light with relativistic effects as clues.

Anthologies

Asimov, Isaac, Ed. *Where Do We Go From Here?* Garden City, NY: Doubleday, 1971. A collection of science-as-subject and plot device stories with explanations of the principles involved by Asimov.

Clarke, Arthur C., Ed. *Time Probe: The Sciences in Science Fiction*. New York: Delacorte Press, 1966. Short stories using science concepts.

Clement, Hal, Ed. *First Flights to the Moon*. Garden City, NY: Doubleday and Co., 1970. Stories about the exploration of the moon with afterwords discussing the scientific concepts involved.

Conklin, Groff, Ed. *Science Fiction Adventures in Dimension*. Stories about topology and dimensions.

Conklin, Groff, Ed. *Science Fiction by Scientists*. New York: Collier Books, 1962. Science fiction stories by scientists, including "Grand Central Terminal" by noted nuclear scientist Leo Szilard.

Preiss, Byron, Ed. *The Microverse*. New York: Bantam Books, 1989. Articles by noted scientists on cells, DNA, subatomic particles, quarks, and quantum theory, followed by SF stories incorporating the concepts.

Preiss, Byron, Ed. *The Planets*. New York: Bantam Books, 1985. A collection of scientific articles, speculative stories, photographs, and illustrations of the planets.

Preiss, Byron, Ed. *The Ultimate Dinosaur*. New York: Bantam Books, 1992. Articles by paleontologists on the latest advances in dinosaur theories with stories incorporating the ideas in fiction.

Preiss, Byron, Ed. *The Universe*. New York: Bantam Books, 1987. Scientific articles and speculative stories on stars, pulsars, black holes, galaxies, quasars, and cosmology.

Schmidt, Stanley, Ed. *Analog Reader's Choice*. A "best of" collection from *Analog* Magazine, the science fiction magazine noted for publishing "hard" science fiction.

Scithers, George, Ed. *Isaac Asimov's Worlds of Science Fiction*.

List B: Provided by James Gunn

Asimov, Isaac. *The Gods Themselves*. His most chemistry-related novel. (Allegedly written in response to a challenge to justify an unacceptable $n°/p^+$ ratio in the atoms of an element—Editor.)

Balzac, Honoré de. *The Philosophers Stone*. 1834. Deals with alchemy.

Bear, Greg. *Queen of Angels*. Good hard-science fiction novel.

Disch, Tom. *Camp Concentration*. A wonder drug heightens intelligence at the expense of death within months.

Gunn, James. *The Dreamers*. Based on chemical memory.

Gunn, James. *The Immortals*. Updated from its original 1962 publication; feature film rights have been purchased by Walt Disney pictures. The story deals with some consequences of mutational improvement in human blood.

Stevenson, Robert L. *Dr. Jekyll and Mr. Hyde*. The classic mad-scientist chemist.

Wells, H. G. *The New Accelerator*. A chemical can speed up metabolism so much that time seems frozen; the story also involves questions about scientific responsibility for discoveries.

List C: Provided by James O'Brien

Bell, H. W., Ed. *Baker Street Studies*. New York: Otto Penzler Books, 1934.

Bunson, M. E. *Encyclopedia Sherlockiana*. New York: Macmillan, 1994.

Coren, M. *Conan Doyle*. London: Bloomsbury Publishing, 1995.

Green, R. L., Ed. *The Uncollected Sherlock Holmes*. London: Penguin Books, 1983.

Holroyd, J. E. *Baker Street By-Ways*. New York: Otto Penzler Books, 1959.

Redmond, C. *A Sherlock Holmes Handbook*. Toronto: Simon & Pierre, 1993.

Ruthman, S., Ed. *The Standard Doyle Company: Christopher Morley on Sherlock Holmes*. New York: Fordham University Press, 1990.

Starrett, V. *The Private Life of Sherlock Holmes*. New York: Pinnacle Books, 1910.

Editor's Supplement

Lovisi, Gary. *Science Fiction Detective Tales*. An overview of these hybrid tales. Paperback from Gryphon Publications, P.O. Box 209, Brooklyn, NY 11228-0209.

Quite a number of titles could be added here, including *The Caves of Steel* by Isaac Asimov (1954), said to have been written in response to a challenge to write a true hybrid of the SF and the whodunnit genres. Asimov's book can be recommended to aficionados of both genres. Its sequel, *The Naked Sun*, likewise qualifies as an example of the combined form. Additional short stories featuring Asimov's detective Wendell Urth, clearly modeled after R. Austin Freeman's fictional scientist-detective Dr. Thorndyke but placed in the future, have been collected in the book *Asimov's Mysteries* (1968).

Asimov also wrote more traditional mysteries. Uniquely pertinent for chemists is his story dealing with the death (or murder?) of a graduate student that takes place in the laboratory of a college chemistry department. It has been frequently republished under the various titles *The Death Dealers*, *A Whiff of Death*, and *Death in the Laboratory*.

Randall Garrett has effectively combined the element of sorcery and sleuthing in a series of stories featuring his "detective" Lord Darcey solving mysteries by Sherlockian deductive powers in an alternate earth where magic is a science. The several books—*Murder and Magic* (shorter stories), *Too Many Magicians* (a novel), and *Lord Darcey Investigates*—have been collected in the single volume *Lord Darcey*.

There is also Gil Hamilton, who is a policeman with some wild mental talents and an operative in a global police force in the future. Several stories, all well done, appear in the book, *The Long Arm of Gil Hamilton* by Larry Niven (1976). Of particular interest is the book's afterword, entitled "The Last Word About SF Detectives", which offers pertinent comments and some well-chosen additional titles.

List D: Provided by Mark Nanny

This bibliography contains works of science fiction that were not included in Nanny's chapter but are worth mentioning. Either these works do not quite fit the scope of the chapter, or they mention very little planetary chemistry, or they are based upon fantasy. There are also a few excellent works listed here that were discovered too late to be included in the chapter.

Venus
Bradbury, Ray. "The Long Rain". 1950.
Pohl, Frederik. *The Annals of the Heechee.* 1987.
Pohl, Frederik. *Heechee Rendezvous.* 1984.
Smith, George O. *Venus Equilateral.* 1975.

Earth
Aldiss, Brian W. *The Long Afternoon of Earth.* 1962.
Ballard, J. G. *The Burning World.* 1964.
Ballard, J. G. *The Drowned World.* 1962.
Bass, T. J. *The Godwhale.* 1974.
Brunner, John. *The Sheep Look Up.* 1972.
Calvino, Italo. *Cosmicomics.* 1968.
Christopher, John. *No Blade of Grass.* 1957.
Clarke, Arthur C. "The Forgotten Enemy." 1953.
Elwood, Roger and Virginia Kidd, Eds. *The Wounded Planet.* 1973.
Herbert, Frank. *The Green Brain.* 1966.
Moore, Ward. *Greener Than You Think.* 1947.

The Moon
Boulle, Pierre, *Garden on the Moon.* 1965.
Budrys, Algis. *Rogue Moon.* 1960.
Clarke, Arthur C. A *Fall of Moondust.* 1961.

Clarke, Arthur C. *A Prelude to Space.* 1954.
Heinlein, Robert A. *The Moon is a Harsh Mistress.* 1966.

Mars

Asimov, Isaac. *The Martian Way.* 1969.
Bear, Greg. *Moving Mars.* 1993.
Bradbury, Ray. "The Blue Bottle." 1950.
Bradbury, Ray. "The Exiles." 1950.
Bradbury, Ray. *The Martian Chronicles.* 1950.
Bradbury, Ray. "The Naming of Names." 1949.
Bradbury, Ray. "The Visitor." 1948.
Clarke, Arthur C. "The Snows of Olympus: A Garden on Mars." 1995.
Clarke, Arthur C. "The Trouble with Time." 1960.
Farmer, Philip Jose. "Jesus on Mars." 1979.
Heinlein, Robert A. *Red Planet.* 1949.
Hipolito, Jane, and Willis E. McNelly, Eds. *Mars, We Love You: Tales of Mars, Men and Martians.* 1971.
Lovelock, James and Michael Allaby. *The Greening of Mars.* 1984.
Stapledon, Olaf. *The First and Last Men.* 1930.

The Asteroid Belt

Anderson, Poul. *Tales of Flying Mountains.* 1970.
Clarke, Arthur C. "Summertime on Icarus." 1960.
Jameson, Malcolm. *Prospectors of Space.* 1940.
Simak, Clifford D. "The Asteroid of Gold." 1932.

Jupiter

Anderson, Poul. "The Snows of Ganymede." 1958.
Benford, Gregory, and Gordon Eklund. *If the Stars Are Gods.* 1977.
Benford, Gregory. *Jupiter Project.* 1975.
Clarke, Arthur C. "Jupiter Five." 1953.
Del Rey, Lester. "Outpost of Jupiter." 1963.

Saturn

Clarke, Arthur C. "Saturn Rising." 1961.

Neptune

Delany, Samuel R. *Triton.* 1976.
Stapledon, Olaf. *The First and Last Men.* 1930.

List E: Provided by Jack Stocker

Anderson, Poul. *Satan's World.* 1968. Deals with mining heavy metals on a rogue planet under hellish conditions; part of a highly developed "future history" by a major "hard science" author.

Brin, David. *Startide Rising.* 1983. (sequel: *The Uplift War,* 1987). Written by a major, newer SF writer, involving alien life, genetic "uplifting", and a galactic library.

Clement, Hal. *Mission of Gravity.* 1954. A classic hard SF story, the background of which involves the beautifully realized, highly detailed description of a complex planet "Mesklin" (including its inhabitants).

Crichton, Michael. *Jurassic Park.* 1996. (and sequel). Lots of biochemical talk and double-talk. The author's *Sphere,* a much less successful science fiction attempt, has some interesting hardware.

Forward, Robert. *Dragon's Egg.* 1981. (sequel: *Starquake*). Considers life on the surface of a neutron star. The author is viewed as a major writer of hard SF.

Hoyle, Fred. *October the First is Too Late.* 1966. A complex novel dealing with coexisting historical-time zones and is based on the many-worlds interpretation of quantum mechanics.

Heinlein, Robert. *Farmer in the Sky.* 1950. Ostensibly one of Heinlein's series for juveniles, almost all of which can be read with pleasure by mature adults, this book deals with a family's trials and tribulations homesteading in the asteroids.

Jensen, William. *Captain Nemo's Battery: Annotations on the Chemistry of Classic Science Fiction.* Monograph. University of Cincinnati, 1996. A very scholarly work that is particularly recommended.

Le Guin, Ursula K. *The Left Hand of Darkness.* 1969. This finely textured, highly literate novel is often accorded the accolade of being among the very finest science fiction novels ever written. Among its virtues, it deals fundamentally with the matter of gender, utilizing a race for which circumstances beyond individual control determine reproductive roles for any one period of time. Strongly recommended.

Pohl, Frederik. *Man Plus.* 1976. Details the bioengineering of humans to live unprotected on Mars.

Sheffield, Charles. *Between the Strokes of Night.* 1985. Considered a "hard science fiction" writer, Sheffield describes a life form in intergalactic space. (*The Black Cloud* by the astronomer Fred Hoyle also did this several decades ago.)

Vance, Jack. *The Blue World.* 1966. Set on a world of water with floating
islands supporting the planet's inhabitants, this story fully describes the
latter's way-of-life, including their interfacing with the planet's unique
ecology.

Surveys and Histories of the Science Fiction Field

Aldiss, Brian W. *Billion Year Spree: The True History of Science Fiction.* Doubleday, 1973. A general, critical survey of the field by a leading writer; a
much-quoted book.
Del Ray, Lester, Ed. *Fantastic Science Fiction Art, 1926–1954.* Ballantine
Books, New York, 1975.
Gunn, James. *Alternate Worlds: The Illustrated History of Science Fiction.*
Prentice Hall, 1975. Reprinted by A&W Visual Library, New York,
softcover, in 1976. It has been described as an "informed study of the
scientific, social, and philosophical climate which brought forth and
shaped science fiction". The reader will find it eminently readable and
generously laced with illustrations, many in color. While not recent, it
is probably still the best book of its sort available.
Gunn, James, Ed. *The Road to Science Fiction.* White Wolf Publishing, 1979.
An excellent four-volume series that consists of Vol. 1, *From Gilgamesh
to Wells;* Vol. 2, *From Wells to Heinlein;* Vol. 3, *From Heinlein to Here;* and
Vol. 4, *From Here to Forever.* In addition to extensive commentary, volume 3, for instance, offers 36 stories/articles that could not, in the editor's view, have been better chosen. (References after chapter 1 provide further details.)
Lundwall, Samuel J. *Science Fiction: What It's All About.* Ace Books, 1971. A
bit elderly but still excellent as an introduction.
Nichols, Peter, Ed. *The Science in Science Fiction.* New York: Crescent Books,
1987. Distributed by Crown Publishers. A thoroughly fascinating book,
critically assessing the implicit/explicit science presumed by the authors
who have published in this field over a broad period of time.

Computers, Cyberworlds, and Nanotechnology

A significant omission in this book is the lack of chapters dealing with science fiction exploitations of the possibilities inherent in computers/cyber-
worlds and nanotechnology. While "high-tech" has long been a significant
component of hard science fiction (as well as in the multitude of modern
thrillers), the ever-expanding horizons of cyberdom, including virtual real-

ity, artificial intelligence, and a long list of additions, often carry a chemical component of a material science or biochemical nature.

Authors who have been admired for their writing in the virtual reality, cyberpunk, and artificial intelligence areas include William Gibson, Bruce Sterling and Wilhelmina Baird, among others.

Developments in science at the nano- or 10^{-9} level, as prefixing a dimension in meters, correspond to atomic diameters; (1 nanometer = 10 Ångstroms). One indication of the level of current interest in nanotechnology can be inferred from the symposium topics at *both* the 1995 *and* the 1996 Robert A. Welch Conferences. Often considered to reflect the hottest broad umbrella areas of current interest and activity, these topics dealt with nanoscale chemistry and were entitled "Nanophase Chemistry" and "Chemistry on the Nanometer Scale", respectively.

The following listings are not clear-cut and have not all been personally evaluated by the editor, who has borrowed generously from an Internet offering (explicitly entitled "Nanotechnology in Science Fiction, Bibliography" Version 1.9, Sept. 21,1995) from anthony@lexis-nexis.com. (http://www.eri). This list rates the stories for their nanotechnology content and includes entries for movies, television, and foreign language offerings.

In general, the specific authors listed below have quite likely contributed additional offerings in these areas, so you might consider their other books not listed here. (Note that, with a few exceptions, the titles all appeared in the 90s.)

Strongly Recommended

Bear, Greg. *Blood Music.* 1985. Often considered the earliest novel to deal explicitly with nanotechnology.

Flynn, Michael. *Nanotech Chronicles.* 1991. (See also *In the Country of the Blind.*)

Gibson, William. *Neuromancer.* 1984 (1994). (See also *Count Zero* and *Burning Chrome* by the same author).

Goonan, Kathleen Ann. *Queen City Jazz.* 1994.

Sterling, Bruce. *Crystal Express.* 1989. A collection of short stories. (See also *Globalhead*, short stories, and *Mirrorshades*—Editor.)

Stephenson, Neil. *The Diamond Age.* 1995. Hugo award winner. (See also *Snow Crash.*)

Williams, Walter Jon. *Aristor.* 1992. (See also *Hard Wired.*)

Other Recommended Titles

Anderson, K. and D. Beason. *Assemblers of Infinity.* 1993.

Baird, Wilhelmina. *Crashcourse.* 1993. (See also *Clipjoint* and *Psykosis.*)
Bova, Ben. *Voyager 3: Star Brothers.* 1990.
Elliot, Elton, Ed. *Nanodreams.* 1992.
Keith, William. *Warstrider.* 1993. (two sequels).
McAuley, J. Paul. *Fairyland.* 1995.
Nagata, Linda. *The Bohr Maker.* 1995
Preuss, Paul. *Human Error.* 1985.

Index

Abandoned, The, 181
Across the Zodiac: The Story of a Wrecked Record, 56
Adventure of the Blue Carbuncle, The, 120
Adventure of the Copper Beeches, The, 114–15
Adventure of the Dancing Men, The, 114
Adventure of the Empty House, The, 115
Adventure of the Six Napoleons, The, 112
Adventures of Sherlock Holmes, The, 108
After Many a Summer Dies the Swan, 27
Aldiss, Brian, 25, 40
All You Zombies, 24
Alternate, The, 181
Amazing Stories, 6, 10, 136
    cover art, 136–37
American Cryptogram Association, 165
American Physical Society Membership Directory, 242
Ammonium, 80
Analog, 126
Analog Science Fact, cover art, 144–45
Analog Science-Fact, Science Fiction, 10
Anderson, Poul, 8, 13, 25, 30
And Now The News..., 27
Andromeda Strain, The, 22–23
Annual Reviews, 18
APA, 15

Arena, 178
Armstrong, Neil, 61
Arnold, John E., 251
Asimov, Isaac, 8, 11, 13, 150
    criticism of Sherlock Holmes, 113–14, 120–21
    life of, 78–81
    Pâté de Foie Gras, 87–88, 211–25
    structure of stories, 21–23, 25, 27, 29–30, 43
    thiotimoline, 82–87, 205–10
Astounding Science Fiction, 6, 10, 80–82, 84, 87, 126, 129
    cover art, 138–43, 146–49
Astounding Stories, 10, 126
At The Rialto, 28

Babylon 5, 175
Baker, Tom, 177
Balfour, Arthur, 49
Ballard, J. G., 28, 31
Battlestar Galactica, 175
Becalmed in Hell, 25
Beep, 24
Before Eden, 61
Bell, Joseph, 108
Bellamy, Ralph, 8
BEM, 15

Berichte, 118
Beryllium, 81–82
Berzelius, Jon Jacob, 119
Bicentennial Man, The, 27
Big Pat Boom, The, 31
Birds Can't Count, 31
Blake's Seven, 175
Blish, James, 24–25, 27
Blood chemistry, 119
Blot, 30
Blue Mars, 73
Blundering Chemist, 120
Body Snatchers, The, 30
Bok, Hannes, 136
Bonstell, Chesley, 14, 136, 140
Boscombe Valley Mystery, The, 110, 112
Bova, Ben, 70
Boy and His Dog, A, 12
Bradbury, Ray, 25, 27
Bradley, Marion Zimmer, 12
Brain of Morbius, The, 191
Brave New World, 4, 8, 45
Bread and Circuses, 193
Brin, David, 22
Brown, Fredric, 24
Bryant, Ed, 28
Buck Rogers, 175
Burroughs, Edgar Rice, 10, 235
By His Bootstraps, 24
Byron, George Gordon, Lord, 37

Call Me Joe, 25
Campbell, John W., Jr., 10, 30, 79, 125–26,
    128–29, 132
Captain Future Magazine, 6
Caravan, T. P., 129–32
Caretaker, The, 194
Carroll, Lewis, 9
Cartmill, Cleve, 26
Case of Identity, A, 110
Cat's Cradle, 173
Caves of Andronzani, The, 191
Certainty for a Future Life on Mars, The, 56
Chains of Command, 184
Chase, The, 180
Chemical and Engineering News, 144, 238
Chemistry, 38, 42, 45–46, 48, 92–102,
    211–25. See also Nuclear chemistry
    of blood, 119
    chirality, 227–30
    in Dr. Who, 173–97

Chemistry—Continued
    industry and, 168
    planetary, 53–75
    in science fiction, 125–34
    in Sherlock Holmes stories, 107–22
    in Star Trek, 155–68, 173–97
    in stories of Asimov, 77–89
    thiotimoline, 205–10
Child By Chronos, 24
Childhood's End, 30
Chronic Argonauts, The, 41–42
Churchill, Winston, 9, 49
Circuit Riders, The, 26
City on the Edge of Forever, The, 188, 193
Clarke, Arthur C., 5, 8, 21–24, 30, 62,
    66–67
Claws of Axos, The, 176
Clement, Hal, 30
Clingerman, Mildred, 31
Code for Sam, A, 30
Code of Honor, 184
Cold, Cold Grave, The, 131
Cold Equations, The, 25
College Chemistry Faculties, 242
Colony in Space, 192
Columbus of Space, A, 56
Comets, 68–69
Compounded Interest, 23
Computers Don't Argue, 27
Conklin, Groff, 22
Conrad, Joseph, 92
Conventions, SF, 16–17
Corrigan
    John, 6
    Joyce, 6
Cosmic Argonauts, 43
Cost of Living, 181, 193
Cover art, 13–14, 137–53
Crichton, Michael, 3, 22
Crick, Francis H. C., 164
Crucifixus Etiam, 58
Curse of Fenric, The, 193

Dangerfield, Rodney, 4
Danse Macabre, 9
Dark Soul of Night, The, 25
Darwin
    Charles, 41
    Erasmus, 36–37
Data's Day, 180
Davy, Sir Humphry, 37

Dawson, Charles, 82
Deadline, 26
Deadly Years, The, 156–57, 159–61, 184, 190
Delany, Samuel, 28
Del Rey, Lester, 30
Desertion, 60
Destiny, 187
Devil in the Dark, The, 178, 194
Diamond Maker, The, 42
Dick, Philip K., 26–27
Dickson, Gordon, 12–13, 27
Dieppe, Carol, 24
Dilemma, 30
Dinosaur Day, 131
Disaster, 184
Do Androids Dream of Electric Sheep?, 27
Dominators, The, 176
Doom of London, The, 55, 57
Doyle, Sir Arthur Conan, 107–8
Dr. Who, 173–78, 189, 191–92, 196
    audience size, 174
    chemistry, 175–78
    portrayal of scientists, 182–84
Dracula, 9
Drumhead, The, 163–64, 180
Dune, 194
Dungeons and Dragons (D & D), 14

Effinger, George Alec, 28
Elaan of Troyius, 178, 184
Electricity, 36
Eliot, T. S., 92
Ellison, Harlan, 11–12, 28
Emanations, 182
Emshweiler, 150
Endochronic Properties of Resublimated Thiotimoline, The, 23, 205–10
Endocrinology, 79
Enlightenment, 191
Escher, M. C., 24
Eternal Adam, The, 40
Ethics, 186
Evolution, 180, 185

Facts in the Case of M. Valdemar, The, 37
Fan clubs, 16
Fans, 14–18
Fantastic Voyage, 25
Farenheit 451, 27
Fariña, Richard, 92

Feynman, Richard, 24
FGOH, 15
FIAWOL, 15
FILKSINGING, 15
Filmer, 45
Final Mission, 184, 192
Final Problem, The, 115, 122
Finlay, Virgil, 136
Finney, Jack, 30
Firstborn, 181
First Flights to the Moon, 21–22
First Men in the Moon, The, 25, 46, 48
Fitzpatrick, R. C., 26
Five Doctors, The, 191
Five Orange Pips, 112
Flash Gordon, 174
Food of the Gods, 48
Forbes, 164
Force of Nature, 186
Forward, Robert, 8
Foster, Natalie, 168
Fourth Law of Robotics, The, 30
Frankenstein, 9, 27, 91
Freas, Kelly, 148
From the Earth to the Moon, 38

GAFIA, 15
Galaxy's Child, 194
Galaxy Science Fiction, 11
    cover art, 150–51
Galileo Seven, The, 178
Game, The, 184
Genesis, 180
Genesis of the Daleks, 183, 192
Genetics, 157–59, 161–62, 164
Geology, planetary, 70–73
Gernsback, Hugo, 7, 17–18, 26, 136
Gerrold, David, 24
Gibson, William, 8, 23
Gilmour, Sir John, 108
"*Gloria Scott*," *The*, 113
Gods Themselves, The, 43, 81
Godwin, Tom, 25
GOH, 15
Golden Apples of the Sun, The, 25
Gravity's Rainbow, 91–102
Green Death, The, 183, 192
Green Mars, 69, 71
Griffin's Egg, 25
Gulliver's Travels, 36
Gunn, Eileen, 28

Halley's Comet, 68
Handbook of Chemistry and Physics, 118
Hand of Fear, The, 177
Happy Solution, 130
Harman, Denham, 159
Harness, Charles, 24
Harrison, Harry, 27, 30
Harrison Bergeron, 27
Hartnell, William, 175
Have Space Suit, Will Travel, 25
Heart of Darkness, 92
Heat Death of the Universe, The, 28
Hector Servadac, 38
Heinlein, Robert, 11, 13, 24–26, 30, 138
Helfand, David J., 22
Hide! Hide! Witch!, 142
Hollow Pursuits, 185
Home Soil, 179, 184
Horizon, 174
Howard, W. E., 59
How Beautiful with Banners, 63
Hubbard, L. Ron, 12
Hunted, The, 185
Huxley
    Aldous, 8, 27, 45
    John, 136
    Thomas H., 41, 43

I, Robot, 30
I Am Legend, 6
Impolex G, 96–97
Inferno, 183
Influence of Trade upon the Form of the Hand,
    with Lithotypes of the Hands of Slaters,
    Sailors, Cork Cutters, Compositors,
    Weavers, and Diamond Polishers, The,
    110
In Search of Wonder, 26
Internet Yellow Pages, The, 17
In the Days of the Comet, 48
In the Hands of the Prophets, 187
Invasion of the Dinosaurs, 196
Invisible Man, The, 43
Island of Dr. Moreau, The, 43

Jetrel, 187
Johnny Mnemonic, 8
Joseph, M. K., 37
Journey in Other Worlds, A, 56
Journey to Babel, 163
Journey to Mars, A, 56

Journey to the Center of the Earth, 38
Jurassic Park, 3, 181

Kahn, David, 163
Keeper of Traken, The, 191
Kepler, Johannes, 35
Kind of Murder, A, 24
King, Stephen, 9, 235
Knight, Damon, 26, 30–31
Kornbluth, C. M., 30
Kraken Wakes, The, 31
Krotons, The, 176

Land of the Giants, 175
Last Man, The, 37
Last Minute, 130
Latham, Philip, 22–23
Lavoisier, Antoine Laurent, 80
Learning Curve, 182
Lederman, Leon, 22
Legacy, 180–81
Leinster, Murray, 26
Lessons, 186
Les Voyages Extraordinaires, 38
Liar, 30
Liberation of Earth, The, 31
Light of Other Days, The, 26
Locher, Dick, 4
Logic, 108–9
Logopolis, 177
London Times, 197
Looking Backward, 8
Loss, The, 184
Lost in Space, 175
Lovecraft, H. P., 7, 9
Lundwall, Sam, 6

MacDonald, Anson, 138
Magazine of Fantasy and Science Fiction,
    The, 11
Magnificent Possession, 80
Mailer, Norman, 92
Make Room, Make Room!, 27
Man Alone, A, 181
Man of Steel, Woman of Kleenex, 23
Man Trap, The, 193
Man Who Folded Himself, The, 24
Man with the Twisted Lip, The, 112
Many Mansions, 24
Markham, Russ, 30
Marooned on Vesta, 25

Mars, 69
Marsh, Sir Edward, 49
Martian Chronicles, The, 25
Martian Odyssey, A, 57
Masterpiece Society, The, 186, 192
Matheson, Richard, 6
Matter of Honor, A, 193–94
Matter of Time, A, 196
Mawdryn Undead, 183, 191
McCaffrey, Anne, 12
McGann, Paul, 175
Measure of a Man, 185
Meeting with Medusa, A, 63, 67
Ménage à Troi, 184
Men and the Mirror, The, 24
Men Like Gods, 8, 49
Merchants of Venus, The, 62
Merck Index, The, 118
Mesmerism, 37
Metamorphosis of Earth, The, 31
Metropolis, 91
Microwave, The, 22
Miri, 190
Mirror, Mirror, 184
Mission Earth, 12
Moissan, Henri, 42
Molecular chains, 99–102
Moon Is Hell, The, 126
Mudd's Women, 188
Mutants, The, 193
Mysterious Island, The, 38–39

Naked Now, The, 166–67, 173, 184
Naked Time, The, 165, 167, 173
Neuromancer, 23
Neutral Zone, The, 191
New Accelerator, The, 46
New Ground, 194
Newton, Isaac, 29
New Yorker, 4
Nightfall, 23, 25
Night Terrors, 181, 184
Nineteen Eighty-Four, 8, 45
Niven, Larry, 8, 13, 22, 24–25
Nor Iron Bars, 25
Nova, 174
Nuclear chemistry, 81, 211–25

Omega Glory, The, 179, 190
Omnilingual, 24
One Sunday in Neptune, 64

On the Beach, 142
*On the Use of Dogs in the Work of the
    Detective,* 110
Orwell, George, 8, 45
Other Worlds Science Fiction, 129
Outer Limits, The, 174

Pal, George, 44
Palmer, Ray, 130
Paradise, 196
Particle Theory, 28
Pâté de Foie Gras, 23, 87–88, 211–25
Patterns of Force, 178
Paul, Frank R., 136
Pedler, Dr. Kit, 175
Piece of the Action, A, 193
Piper, H. Beam, 24
Planetary science
    chemistry, 53–75
    geology, 70–73
    significant events in, 55, 58, 62, 65
Planet of Doubt, The, 60
Planet of Evil, 183
Planet of the Daleks, The, 176
Planet of the Giants, 175, 183, 192
Planets, The, 22
Planet Stories, 135
Plato's Cave, 30
Plato's Stepchildren, 188
Plattner Story, The, 43
Plunge into Space, A, 56
Poe, Edgar Allan, 9, 37
Pohl, Frederik, xxi, 62
Politics, 26
Preiss, Byron, 22
Price, The, 186
Princess of Mars, A, 56
Pronzini, Bill, 11
Prott, 31
Puppet Masters, The, 30
Pynchon, Thomas, 91–92

Quality of Life, The, 186
Quartermass, 174

Ralph 124C41+, 26
Recommended reading
    anthologies, 276–77
    history, 282
    novels, 248, 276–78
    short stories, 271–75

Recommended reading—*Continued*
    surveys, 282
Recommended viewing
    movies, 247
    television, 247
Red-Headed League, The, 110
Red Mars, 23, 69, 71
Red Planet, 25
Redshift Rendezvous, 24
Remarkable Case of Davidson's Eyes, The, 42
Remember Me, 184
Remembrance of the Daleks, 183
Rendezvous with Rama, 23
Requiem for Methuselah, 190
Resurrection of the Daleks, 183
Rest of the Robots, The, 30
Revenge of the Cybermen, 177
Reynolds, Mack, 23
Rice, Anne, 12
Ringworld, 22–23
Robinson, Kim Stanley, 23, 25
Robot, 193
Robots of Death, 183
Rocklynne, Ross, 24
Rogers, Buck, xxi
Roscoe, Henry Enfield, 117
Rose, Mark, 132
Ross, Marvin, 68
Royale, The, 180
Rubner, Max, 160
Runaround, 29, 80
Rutherford, Ernest, 49

Sagan, Carl, xxi
Salinger, J. D., 92
Sands of Mars, The, 58
SCA, 15
Schiller, Friedrich, 81
Schisms, 181
Schwarzschild Radius, 28
Science. *See* Chemistry; Genetics;
    Geology; Nuclear chemistry;
    Planetary science
Science fiction
    classics, 35–51
    conventions, 16–17
    cover art, 13–14, 135–51
    definition, 5–9, 35
    in education, 233–39, 241–49, 251–67
    fans, 14–18
    history of genre, 10–13
    media, 13–14

Science fiction—*Continued*
    planetary chemistry in, 53–75
    popularity of, 3–4
    respectability of, 4–5
    science
        as basis for SF, 29–31, 125–34
        in Dr. Who, 173–97
        in *Gravity's Rainbow,* 91–102
        as metaphor in SF, 26–29
        as SF plot device, 24–25
        in SF stories of Asimov, 77–89
        as SF subject, 21–24
        in Sherlock Holmes stories, 107–22
        in Star Trek, 155–68, 173–97
    scientists in, 182–87
    space opera, 57–60
    temporal chirality spoof, 227–30
    themes of, 13
    via Internet, 17–18
Science Fiction: What It's All About, 6
Science Fiction by Scientists, 22
Science in Cinema, 241
Science Schools Journal, 41
Scientific method, 109–12
Scientists in science fiction, 182–87
Screwfly Solution, The, 28, 30
Seeds of Death, The, 176
Sensorites, The, 175
SFTPOBEMOCOSFM, 15
Shada, 183
Shape of Things To Come, The, 25
Shaw, Bob, 26
Sheldon, Raccoona, 28, 30
Shelley, Mary, 9, 27, 37
Short History of Chemistry, A, 77
Schrödinger's Kitten, 28
Sidon in the Mirror, The, 25
Sign of Four, The, 107–10, 112–13, 118
Silent Spring, 192
Silicon Avatar, 186
Silly Season, The, 30
Silverberg, Robert, 24, 59, 80
Sins of the Father, 194
Slight Case of Sunstroke, A, 24
Smith
    Clark Ashton, 31
    E. E., 6, 12
SMOF, 15
Soddy, Fredrick, 49
Solar system. *See also* Planetary science
    Europa, 66–67
    Jupiter, 53, 60, 65–67

Solar system—*Continued*
  Mars, 54–56, 57–59, 61, 65, 69–73, 79
  Mercury, 54, 58–59, 80
  Moon, the, 38, 54, 61, 126–29
  Neptune, 64–65, 68
  Pluto, 64
  Saturn, 60, 63, 65, 73
  Uranus, 60, 65, 68
  Venus, 54–56, 61–63, 64–65
Somnium, 35
Space 1999, 175
Space, Time and Education, 146, 251–67
Space opera, 57–61
Space Precinct, 175
Space Seed, 178
Special Delivery, 24
Sphere, 3
St. Clair, Margaret, 31
Stable Strategies for Middle Management,
  28
Star Begotten, 49
Startling Stories, 6
Star Trek, 173–74
  audience size, 174
Star Trek: Deep Space Nine, 12, 174, 196
  chemistry, 181
  portrayal of scientists, 187
Star Trek: The Next Generation, 12, 17, 21,
  157–59, 161–66, 173, 189–91,
  193–95, 246
  chemistry, 179–81
  portrayal of scientists, 184–86
Star Trek: The Original Series, 5, 11–12, 14,
  155–68, 178, 188–90, 193–94, 246
  chemistry, 178–79
  portrayal of scientists, 184
Star Trek: Voyager, 12, 174, 181, 194
  chemistry, 182
  portrayal of scientists, 187
Star Trek Generations, 184
Star Trek IV, 194
Star Trek VI, 194
Star Wars, 5, 8, 21
Stevenson, Robert Louis, 40
STF (Scientifiction), 9
Stith, John, 24
Stocker, Jack, 133
Stoker, Bram, 9
Stones of Blood, The, 177
Stoppard, Tom, 92
Story of Days to Come, A, 45
Strand, The, 115

Strange Case of Dr. Jekyll and Mr. Hyde, The,
  40
Stuart, Don A., 126
Study in Scarlet, A, 112–13, 121
Study of Tattoo Marks, A, 110
Sturgeon, Theodore, 4, 27
Sub Rosa, 180
Sucker Bait, 81
Suddenly Human, 180
Sunrise on Mercury, 59
Super Science Stories, 79
Surface Tension, 27
Suspicions, 186
Swanwick, Michael, 25
Swift, Jonathan, 36

Tale of the Twentieth Century, A, 41
TANSTAAFL, 15
Tenn, William, 31
Thine Own Self, 187, 192
Thiotimoline, 82–87, 205–10
Thiotimoline to the Stars, 87
Third Law, The, 30
Thirty-Sevens, The, 182
Tholian Web, The, 184
Thrilling Wonder Stories, 6, 10
Thunderbirds, 175
Time and the Rani, 178
Time for the Stars, 24
Time Machine, The, 41–42
Time Probe: The Sciences in Science Fiction, 22
Time's Arrow, 181
Times Higher, 197
Time Squared, 194
Tin Man, 194
Titan, 25
Titan, 63, 73
Tolkien, J. R. R., 235
To Mars via the Moon, 56
Tomb of the Cybermen, 183
Tomorrow's Worlds, 59
Too Short a Season, 191
To Serve Man, 30
Total Recall, 21
To Venus in Five Seconds, 56
*Tracing of Footsteps with Some Remarks upon
  the Use of Plaster of Paris as a Preserver
  of Impresses, The*, 110
Transfigurations, 180
Trefil, James, 91
Trial of a Time Lord-Terror of the Vervoids,
  183

Troughton, Patrick, 176
True Q, 193
Truth About Pyecraft, The, 48
Tubb, E. C., 12
Tunnel in the Sky, 26
Twenty Thousand Leagues Under the Sea, 38–39
Twilight Zone, The, 174
Two Doctors, The, 183
Two Planets, 56
Two Thousand and One: A Space Odyssey, 5, 11, 14
Two Thousand and Ten: Odyssey Two, 65
Two Thousand Sixty-One: Odyssey Three, 65, 68
Typewriter and Its Relation to Crime, The, 110

Ultimate Dinosaur, The, 22
Universe, The, 22
Unnatural Selection, 157–59, 161–63, 180, 185, 192
Unparalleled Adventure of One Hans Pfaal, The, 38
Up the Long Ladder, 185

Varley, John, 25
Vengeance Factor, The, 180
Verne, Jules, 8, 38–39, 47
Visions, 4
Vogt, A. E. van, 11
Vonnegut, Kurt, 5, 11, 27

Wait It Out, 64
Waldo, 26, 138

Wallingford, Lee, 24
War of the Worlds, The, 30, 44–45, 136
Watchtowers, The, 31
Waveries, The, 24
Weapon, The, 79
Weird Tales, 7
We'll Always Have Paris, 185
Welles, Orson, 44
Wells, Herbert George, 8, 25, 30, 136
   life of, 40–50
Weyl, Theodore, 117
What Might Have Been, 9
When the Bough Breaks, 194, 197
When the Sleeper Wakes, 45
Where Do We Go From Here?, 22
Where No Man Has Gone Before, 178
Where No One Has Gone Before, 184, 194
Who Goes There?, 30, 126
Willis, Connie, 18, 25, 28, 30
Willstäter, Will, 98
Wonderful Adventures on Venus, 56
Wonder Stories, 10
World Set Free, The, 49
Wyndham, John, 31

Xi Effect, The, 23
X-rays, 44

Yehudi Principle, The, 24

Zoline, Pamela, 28

Bestsellers from ACS Books

The ACS Style Guide: A Manual for Authors and Editors (2nd Edition)
Edited by Janet S. Dodd
470 pp; clothbound ISBN 0–8412–3461–2; paperback ISBN 0–8412–3462–0

Writing the Laboratory Notebook
By Howard M. Kanare
145 pp; clothbound ISBN 0–8412–0906–5; paperback ISBN 0–8412–0933–2

Career Transitions for Chemists
By Dorothy P. Rodmann, Donald D. Bly, Frederick H. Owens, and Anne-Claire
Anderson
240 pp; clothbound ISBN 0–8412–3052–8; paperback ISBN 0–8412–3038–2

Chemical Activities (student and teacher editions)
By Christie L. Borgford and Lee R. Summerlin
330 pp; spiralbound ISBN 0–8412–1417–4; teacher edition, ISBN 0–8412–1416–6

Chemical Demonstrations: A Sourcebook for Teachers, Volumes 1 and 2, Second
Edition
Volume 1 by Lee R. Summerlin and James L. Ealy, Jr.
198 pp; spiralbound ISBN 0–8412–1481–6
Volume 2 by Lee R. Summerlin, Christie L. Borgford, and Julie B. Ealy
234 pp; spiralbound ISBN 0–8412–1535–9

The Internet: A Guide for Chemists
Edited by Steven M. Bachrach
360 pp; clothbound ISBN 0–8412–3223–7; paperback ISBN 0–8412–3224–5

Laboratory Waste Management: A Guidebook
ACS Task Force on Laboratory Waste Management
250 pp; clothbound ISBN 0–8412–2735–7; paperback ISBN 0–8412–2849–3

Reagent Chemicals, Eighth Edition
700 pp; clothbound ISBN 0–8412–2502–8

Good Laboratory Practice Standards: Applications for Field and Laboratory Studies
Edited by Willa Y. Garner, Maureen S. Barge, and James P. Ussary
571 pp; clothbound ISBN 0–8412–2192–8

For further information contact:
Order Department
Oxford University Press
2001 Evans Road
Cary, NC 27513
Phone: 1-800-445-9714 or 919-677-0977

Highlights from ACS Books

Desk Reference of Functional Polymers: Syntheses and Applications
Reza Arshady, Editor
832 pages, clothbound, ISBN 0–8412–3469–8

Chemical Engineering for Chemists
Richard G. Griskey
352 pages, clothbound, ISBN 0–8412–2215–0

Controlled Drug Delivery: Challenges and Strategies
Kinam Park, Editor
720 pages, clothbound, ISBN 0–8412–3470–1

Chemistry Today and Tomorrow: The Central, Useful, and Creative Science
Ronald Breslow
144 pages, paperbound, ISBN 0–8412–3460–4

Eilhard Mitscherlich: Prince of Prussian Chemistry
Hans-Werner Schutt
Co-published with the Chemical Heritage Foundation
256 pages, clothbound, ISBN 0–8412–3345–4

Chiral Separations: Applications and Technology
Satinder Ahuja, Editor
368 pages, clothbound, ISBN 0–8412–3407–8

Molecular Diversity and Combinatorial Chemistry: Libraries and Drug Discovery
Irwin M. Chaiken and Kim D. Janda, Editors
336 pages, clothbound, ISBN 0–8412–3450–7

A Lifetime of Synergy with Theory and Experiment
Andrew Streitwieser, Jr.
320 pages, clothbound, ISBN 0–8412–1836–6

Chemical Research Faculties, An International Directory
1,300 pages, clothbound, ISBN 0–8412–3301–2

For further information contact:
Order Department
Oxford University Press
2001 Evans Road
Cary, NC 27513
Phone: 1-800-445-9714 or 919-677-0977
Fax: 919-677-1303